Who Speaks for Earth?

Who Speaks for Earth?

Based on the series of distinguished lectures sponsored by International Institute for Environmental Affairs in cooperation with Population Institute held in Stockholm concurrently with the United Nations Conference on the Human Environment, June 1972

BARBARA WARD

RENÉ DUBOS

THOR HEYERDAHL

GUNNAR MYRDAL

CARMEN MIRÓ

LORD ZUCKERMAN

AURELIO PECCEI

Edited by Maurice F. Strong

W · W · NORTON & COMPANY · INC ·
NEW YORK

FIRST EDITION

Library of Congress Cataloging in Publication Data
Main entry under title:

Who speaks for earth?

Seven lectures sponsored by the International Institute for Environmental Affairs during the United Nations Conference on the Human Environment held in Stockholm in June 1972.

    1. Human ecology—Addresses, essays, lectures.
    2. Population—Addresses, essays, lectures.
    3. Pollution—Addresses, essays, lectures.
    I. International Institute for Environmental Affairs.

GF8.W49   301.31   72–11726
ISBN 0–393–06392–5

Contents

THE INTERNATIONAL INSTITUTE FOR ENVIRONMENTAL AF-
FAIRS is an independent, nongovernmental, nonprofit organization. It was established in January 1971, to serve as
a clearinghouse and catalyst for action in environmental
affairs. Its publications, meetings, and workshops are designed to assist public and private institutions by providing a multidisciplinary, international perspective to efforts
to enhance the human environment. It is headed by an
international board of directors and is assembling an international staff. Plans call for several offices of the International Institute to be established throughout the world.

Robert O. Anderson

The Honorable Roy Jenkins } Co-chairmen

Jack Raymond   President

United Nations Plaza at 345 East 46th Street
New York, N.Y. 10017

THE POPULATION INSTITUTE, independent, nongovernmental, nonprofit, helps peoples and governments examine
the relationship of their population growth and resource
consumption to the future well-being of themselves, their
environment, and human society. The Population Institute provided the financial support for the Distinguished
Lecture Series.

Dr. George W. Crawford   Chairman

Rodney Shaw   President

100 Maryland Avenue, N.E.
Washington, D.C. 20002

7

Preface

The United Nations Conference on the Human Environment, held in Stockholm in June 1972, was a conference of governments—110 in all. This first effort to cope with environmental problems on a comprehensive and global basis was marked by considerable achievement. But as must inevitably be the case in a political meeting, some issues were avoided and others treated in a highly partisan manner. Moreover, the delegates had to plod through scores of recommendations prepared in advance, so that the debate was inescapably narrow in focus and often devoted to tedious details of phraseology.

To enhance the work of the Conference, the International Institute for Environmental Affairs, with the support of the Population Institute, sponsored a Distinguished Lecture Series by men and women of international reputation who were invited to speak their minds, free of national interest. Each lecture was devoted to a major environmental issue, including some which—like population—were treated only tangentially by the Conference.

In order to broaden the geographical representation and

the points of view, each of the principal speakers was introduced by one or more persons from a different discipline and cultural tradition. The introducers were invited to say something of substance on the topic in hand, in addition to performing the formality of presenting the principal speaker. These introductions have been greatly shortened in this published version, but the principal addresses are printed in full.

Because of their unofficial connection with the Stockholm Conference as co-authors of the background document *Only One Earth: The Care and Maintenance of a Small Planet,* Barbara Ward and René Dubos spoke in the Folkets Hus, where the plenary sessions of the Conference were held. They were introduced by Maurice Strong, of Canada, Secretary-General of the Conference. The other lectures in the series were held in the Mirror Ballroom of the Grand Hotel.

Philip W. Quigg

Director of Publications
International Institute for Environmental Affairs

About the Authors

Maurice F. Strong, Secretary-General, United Nations Conference on the Human Environment.

Barbara Ward (Lady Jackson) is Albert Schweitzer Professor of International Economic Development at Columbia University. With René Dubos, she is co-author of *Only One Earth: The Care and Maintenance of a Small Planet*, an unofficial report commissioned by the Secretary-General of the United Nations Conference on the Human Environment. Her other works include *Five Ideas That Change the World; India and the West; The Rich Nations and the Poor Nations; Nationalism and Ideology;* and *The Lopsided World*. She is a member of the Pontifical Commission for Justice and Peace.

René Dubos, co-author with Barbara Ward of *Only One Earth*, has achieved distinction as a scientific humanist with such books as *Man Adapting; Man, Medicine and Environment; So Human an Animal*, which won a Pulitzer Prize in 1969; *Reason Awake: Science for Man;* and *A God Within*. A microbiologist and experimental pathologist, he was associated with Rockefeller University for

11

forty-four years. He recently retired but continues writing and lecturing. In 1972 he was the first recipient of the Institut de la Vie Award.

> Letitia E. Obeng, Director, Institute of Aquatic Biology, Accra, Ghana.

Thor Heyerdahl, the anthropologist and explorer, has led numerous explorations which have electrified the world. These include the *Kon-Tiki* expedition from Peru to Polynesia, an expedition to Easter Island, and Atlantic voyages in papyrus boats *Ra I* and *Ra II*. His books include *Kon-Tiki Expedition* and *Aku-Aku: The Secret of Easter Island,* which have been translated into many languages. Mr. Heyerdahl is Vice President of the World Association of World Federalists.

> Saburo Okita, President, Japan Economic Research Center.

Gunnar Myrdal is Professor of International Economy at the Institute for International Economic Studies, University of Stockholm. From 1947 to 1957 he served as Executive Secretary of the United Nations Economic Commission for Europe. At present he is revising his landmark study, *An American Dilemma: The Negro Problem and Modern Democracy.* His other major work is *Asian Drama: An Inquiry into the Poverty of Nations.*

> Avabai Wadia, President, Family Planning Association of India.

Carmen Miró is Director of the Center of Latin American Demographic Studies which is located in Santiago, Chile. A statistician and demographer, she earlier served as the Director of Statistics and Census in Panama.

Chief Adebo, former Executive Director, United Nations Institute for Training and Research.

George Baranescu, President, Commission of Environment, Romanian Academy.

Lord Zuckerman until recently served as Chief Scientific Adviser to the British Government. An anatomist, he is Honorary Secretary of the Zoological Society of London. He is the author of *The Social Life of Monkeys and Apes, A New System of Anatomy;* and *The Scientists and War.*

Ro Chung-Hyun, Director, Institute of Urban Studies and Development, Yonsei University, South Korea.

Aurelio Peccei, President and a founder-member of the Club of Rome, is affiliated with Fiat and Olivetti and manages a consulting firm for economic and engineering development, Italconsult. He is Chairman of the Economic Committee of the Atlantic Institute.

Who Speaks for Earth?

Only One Earth
Introduction by Maurice F. Strong

From the very beginnings of our preparations, it was clear that governments intended the Stockholm Conference to be action-oriented. At the same time, it appeared necessary to provide a conceptual framework for this immensely complex subject for which we have coined the term, human environment. We wanted it to be not merely in a technical setting but in a broad social and political and moral context and in historical perspective.

This, of course, was not a task we felt should be undertaken by a conference secretariat. Bureaucracies seldom produce the kinds of inspiration for which we were hoping. We felt instead that we should reach out to the world's leading thinkers, scientists, historians, jurists, humanists, philosophers, and others for the intellectual perspective against which to view our action plan.

In the final analysis, of course, the judgments concerning what went into the conceptual framework had to come from Professor Dubos and Lady Jackson. The end product is a world report on the state of the human environment in the form of a book entitled, *Only One Earth: The Care and Maintenance of a Small Planet,* made available to all delegations to the conference and to hundreds of thousands of people throughout the world.

It is this unique report which provides the key-note of this special meeting opening the Distinguished Lecture Series.

Speech for Stockholm

By BARBARA WARD

When my dear colleague Professor René Dubos and I were asked to prepare a "conceptual framework" for the Stockholm Conference and set out upon consultations with our distinguished experts all round the world, believe me, we did *not* acquire a conceptual framework at the end of our intellectual journey. What happened was something much more like standing under Niagara, under a cataract of ideas and concepts and contradictions and visions. All this made it quite clear that this is a time when people's ideas about the planet they live in and about the way they can survive on it are changing in an absolutely torrential fashion. Even when our experts were in fact reacting with what one might call an old response, they knew they had to frame it in a new way. Even when they were presenting what they thought to be old and established facts, these had somehow to be freshly presented. This turbulence of ideas—old and new—I take to be the outward signs of a vast inner change in man's interior or imaginative environment. And in all history there can be no greater change than this.

I think Stockholm has come at one of the moments in the affairs of man when people begin radically to reconsider how they have to look at their life on earth, how they have to discover what sense their existence makes to them. It is for this reason that the Stockholm Conference can be a moment of incredible excitement, of fresh ideas and new beginnings. It is only when people begin to shake loose from their preconceptions, from the ideas that have dominated them, that we begin to receive a sense of opening, a sense of vision, a sense of the kind of new directions which, I think we would all agree, our poor old planet most desperately needs.

Such times of re-thinking do recur in history. At the end of the civil wars in China 2,000 years ago, before the setting up of the great centralized Han Dynasty, there was an immense intellectual ferment in this most ancient of civilizations—a ferment from which emerged the powerful synthesis of Confucius. More recently our forefathers went through the "Copernician revolution" in which people had almost literally to turn their minds upside down and discover that the sun in fact did not go around them, but they went around the sun. In their passionate resistance to the idea, we can see a terrible sense of vertigo. It was as though they hardly knew where they were anymore. Cosy flat-earthers had to feel the horror of discovering that the planet is a precarious sphere. The followers of Ptolemy had to displace the earth and put the sun at the center of their system. Such changes of perspective shook people to the roots of their being. Well, that is the sort of time we live in now. We too are in one of those times of vertigo. We too live in an epoch in which the solid ground of our preconceived ideas shakes daily under our uncertain feet.

If René Dubos and I were to describe to you the full scale of ideas and visions that came to us from our consultants, we would keep you here until midnight. Incidentally, you must be gluttons for oratory to be here at all, and I must say, I congratulate you. But do not be alarmed. We have no pretensions to exhaustiveness—or exhaustion. All I think I can usefully do is to pick out three particular instances which seem to me to show that people are beginning to think in a quite new way about their planetary existence.

I doubt if I need talk at any great length about the first of these instances—the risk of irreversible planetary damage. You have followed the day's debates and heard of the risk from many sources. Indeed, one of the fascinating things about the present moment is the speed with which truth is moving toward platitude. There are ideas and concepts which, when I wrote them in our preliminary draft last year, made me wonder how far out I could be. Yet today Ministers of the Crown are saying them and that is surely about as far in as you can get. So let us not linger too long on the newly recognized fact that our total natural system—in the biosphere of air and soil and water—could be irretrievably upset by man's activities. Of course, we have always known that we could chip away at it—here a forest destroyed, there a desert spreading. But the idea that the two great reservoirs of life, our airs and oceans, could enter on a general trend toward mortal damage—that is completely new.

In today's debate, as no doubt you noticed, delegates talked above all of the vulnerability of the oceans. Yet only a year ago, this was an entirely new idea. Now it is a *lieu*

21

commun, a near-platitude, and let us thank God for that. The progress toward truism means that the new ideas are penetrating human consciousness with incredible rapidity. They are acquiring the pervasive and mythical quality of accepted popular wisdom and folklore. And if I may, as a Britisher, recall a piece of our own folklore, I could remind you of the courtiers of King Canute. They told him to set up his throne by the North Sea—a very cold, disagreeable sea, I can tell you from personal experience. Then the king was supposed to tell the North Sea tides to stay put. Of course, they did no such thing and the king got his feet wet. Today we are just beginning to learn that we have had something of the same Canutish attitude. We have virtually told the oceans to stay the same, to carry on "their priestlike task of pure ablution" while loading them up with every kind of junk. We have assumed we could use them as giant sewers and still keep their "priestlike" functions intact.

But now they are beginning to give us their first warning signals. In the last analysis they make up a system without an outlet. Everything that is flushed out from man's activities tends to end irreversibly in them. How much can they carry? We do not know. How much of their tolerance is already used up? Again, we do not know. But our old self-confidence is fading. We *can* murder the biosphere. This realization is the first element I would pick out in our revolutionary ferment of ideas.

The second I would suggest is the beginnings of doubts about the economics of growth. These are, of course, recent misgivings. I myself grew up at a time of massive nongrowth or wholesale depression and unemployment. In a sense, the whole fabulous market boom of the last twenty-five years had been designed to give enough stimulus to the economy to prevent any future depressions. But the

22

underlying techniques of this stimulus have been based upon a veiled assumption. We have taken the line that so long as the rich improve their material standards as rapidly as possible, so long as we reward the most successful people as highly as possible—from the third home to the fourth swimming pool to the fifth car—then somehow their standard-setting affluence will drag everyone else up behind. It could be called follow-my-leader economics, if you like, or "trickle-down economics."

This approach has worked up to a point in the developed world for two reasons. Coming first to the new technological order, we could lay our hands on the world's vast supply of unexhausted resources, in particular, on all the still-unoccupied temperate farm lands. The second reason is the acceptance of a measure of redistributive justice inside society. We have social instruments of transfer—taxation, welfare, insurance. We do in fact take something from the clever and the rich and the healthy and the strong to give to those in greater need. Even so, "trickle-down" methods leave pockets of desperate and hopeless misery.

But in the world at large, the technique is not working. In the first place, with the flow of aid reduced to a pitiful 0.3 percent of gross national product and trade barriers maintained on every side, there are no instruments of redistributive justice. And now we are faced with a new possibility. No quick bonanza awaits the ex-colonial countries. On the contrary, the assumption that we made until the day before yesterday—that the resources of the planet were virtually unlimited and could provide so much wealth at the top that in the end, the very last sharecropper could get a square meal—that assumption is beginning to be questioned. Ultimately our planet is a finite system. Ultimately its resources run out. Nobody has decided precisely what the limits may be. We know that we can put any number of figures into a computer—of population, of demand, of available resources, of pollution—and come up with the answers that match the figures put in in the first place. This leaves large areas of uncertainty. Shall we collapse by 2010 or 2050, or, by a longer reach of the imagination, the year 3000, or by further centuries still?

What is certain is that if we take an annual 2.5 percent increase in population roughly on to the tercentenary of Thomas Malthus's birth, the world's population could be increasing by a billion a year. And that is clearly impossible. I cite this as simply one of the ultimate limits. There are others—of resource use, of the biosphere's carrying capacity—which tell us unmistakably that we have to learn to achieve a more modest use of resources with a redefinition of living standards toward less consumptive joys. All these are true limits. But I am not sure that the most pressing limit of all may not prove to be the patience with which two-thirds of humanity are prepared to accept the present consequences of trickle-down economics—which is that they stay poor now while we grow richer and may be asked to stay poor forever so that the world's carrying capacity may remain equal to providing incomes of $10,000 a year for developed peoples without precipitating ecological disaster. In a limited biosphere with unlimited material ambitions built into the life style of the developed peoples, this is where we are tending, unless the poor stay very poor indeed. And frankly I doubt if they will accept a world society which is so hopelessly lopsided. Their patience will blow up in anarchy before the biosphere reaches the point of no return.

This chance of worldwide revolt brings me to the third great change in our contemporary thinking. It is straightforward enough. Indeed we have had suspicions about it for some time—at least since World War I. This is the suspicion that we cannot run a functioning planetary society on the totally irresponsible sovereignty of a hundred and twenty different governments. It simply cannot be done.

Now the limitations of sane nationalism may be perfectly clear. But the matter is not so simple. Suppose that you achieved your nationhood just fifteen years ago after two hundred years of colonial control. You would not thank your colonial masters if they now waltzed in to tell you that: "Too bad, nationalism is out of date." We have to have a certain sense of reality about this. Those for whom nationality is the most precious and the most recent symbol of dignity and of full participation in the human community are not likely to feel straightaway the same reservations about the nation which are appropriate to our tired, quarrelsome old nations in Europe, who have been fighting each other almost without ceasing for the last six hundred years. You could in fact say that René Dubos and I spring from national stocks which represent the toughest proponents of learning nationalism by killing off the others. René beginning in France and I beginning in England have perhaps demonstrated a miracle of collaboration in our *Only One Earth,* for certainly our two nations fought each other ferociously and singlemindedly into national identity over at least four centuries. Think of what Shakespeare had to say about these wars: "If death were French, death would have died today." There is a certain zest in our Bard when he speaks of England as:

> This precious stone set in a silver sea
> Which serves it in the office of a wall
> Or as a moat defensive to a house
> Against the envy of less happier lands

—I like that—"less happier lands"!

If now the Europeans are beginning to feel they should

transcend nationalism, the developing peoples could well say that it has taken a long time for the penny to drop! Yet today within this European complex of the oldest and in some ways, most bloody-minded of nations, the concept does grow that mankind must go beyond the limits of purely national government and begin to find out what the "post-national community" is like. I hope that this will prove the foremost vocation of a new Europe. And I hope too that by preserving our cultural and regional varieties within a wider scheme of effective loyalty and decision-making we begin to discover what a necessary transcendence of the nation state really means. It cannot, must not mean a suppression of all variety and a civilization so standardized that we all end up hideously the same. We cannot shirk this risk. Take one aspect of it that is especially alarming in an urbanizing world. Modern architecture appears to me to be absolutely identical and almost equally awful wherever it occurs. Those up-ended waffle-irons, those vertical concentration camps lightly decorated with tasteful gas ovens are the same from Santiago de Chile to Tokyo Bay. When we talk of transcending nationalism, it cannot be to achieve planetary uniformity. It is something much more complex, human, and loveable. It must be based on forms of cooperation which respect local autonomy, which respect diversity, yet build the essential unities of our new global society.

At this point, the natural reaction may be to say: We have never heard such utopian pollyanna. How can we possibly believe that anyone is going to behave in all these new ways? The customs, the habits of millennia are behind the drives of separate sovereignty. We are perfectly used to the idea that we shall continue to be rapaciously in pursuit of the last available dollar or pound or yen we can get our hands on. We have no intention of reckoning the cost of national arrogance and economic greed in terms of the damage done to our shared and inescapably interdependent biosphere. In any case, we argue, the oceans, the resources, the energy reserves, heavens, they are still so vast! Custom will carry us along quite safely. There have been panics before. We have survived them. In any case, who wants to panic when the alternative is to believe that our national income is going to go on and on and up and up and people will get that second car and third television set which represent so much of their sense of what the good life is all about?

So it is surely the brashest, most irrational optimism to believe that developed societies will re-think their concepts of a standard of living. We may talk about less consumptive joys and having more lovely recitals by Birgit Nilson and more opera and education and open spaces and parks and sports—all the enjoyments which can be satisfied without harming the biosphere. But such a switch means that an awful lot of people are not going to get what they feel they want. There are sacrifices involved in a stabilized world with much more stabilized incomes. Can any one suppose that our hungry nations, rich and poor, are ready for anything of the sort?

And another obstacle stands, I think, in the way of hope. On this, alas, I speak perhaps with more expertise than anyone in the world. This is the danger that you can say a thing so often that you begin to think it has happened. For instance, we have been talking about the developing world, about rich nations and poor nations, for twenty years. We know all about the development gap. We constantly remind ourselves that it is growing wider. Yet today we are doing less than ever about it. What possible guarantee have we that today's splendid, ringing, wonderful platitudes about environment and planetary risk and the biosphere are not going to go the same way?

No doubt our governments will now sign the Environmental Declaration. They will talk of looking after the oceans. They may even suggest that they could take a new look at the redistribution of world resources. But what guarantees are there that any of these new approaches and promises to pursue wiser policies for the planetary environment will prove more effective than earlier promises to do better about planetary poverty and need? The same risk is there—to mesmerize ourselves by saying a thing is going to happen and then somehow kid ourselves that it has actually occurred. But it has *not*. It is the very same mixture as before and just as our rhetoric has rolled over the facts of starving children and festering cities, it can now roll on over the realities of dying oceans and a wasted earth.

Is the conclusion, then, that between the force of settled habit, existing institutions, and established interests on the one hand, and the bemusing powers of platitude on the other, nothing much is going to change and that we and our planet are rushing on toward that crisis point where population, pollution, resource exhaustion and social alienation combine in some ultimate explosion?

The short answer is, of course, that the outcome depends upon the action and awareness of this planet's citizens. The needed changes in direction can only happen if there are enough people in the world who have a new vision and are prepared to work and sacrifice and persuade and exhort and stay with it, not only when they are young but when they are—as I increasingly feel—dead old and infinitely decrepit. In other words, right through and right on!

To me the great importance of Stockholm is the fantastic number of people who have come to this place, under their own steam, paying their own expenses, and determined to participate because they want to play a role in saving the planet they love. Now, of course, if we want to

28

accomplish so vast a task, we have to find ways of being very specific. We have to become absolutely committed lobbies for concrete acts of policy designed to safeguard our single earth. And we *must* stay with the task. It is so easy to go for one cause—bald eagles—one year, one cause —ghetto dwellings—the next, another cause—the whales— the next, and so on and so on. And all the time, as someone said—probably Adam Smith—the interests never sleep. I do not, of course, mean that all these issues are not vital. But the lobbies must stay with each of them, cooperate in a general movement, reinforce each other's efforts, allot tasks, share offensives and make their own action as sustained, determined and indeed aggressive as the attitudes and institutions they have to reform. If Stockholm is to become the start of something new in our comprehension of our planetary predicament, we simply have to set to work on this active, realistic, and effective citizens' campaign.

Some will no doubt say that this is really the final argument for pessimism. How can we suppose such dedication will ever appear and be sufficiently sustained? But on the contrary, I find the need for new citizen action a reason for hope. This is because, as I said at the beginning, we are entering a new epoch of ferment and ideas. It can be one of those times in human history when an extraordinary change of direction takes place in what people are ready for and in what people want. Let me give you just one example of such a profound redirection of human attention, one which seems to me the most extraordinary and possibly the most relevant in man's history until our own day. Historians tell us that this revolution in thought occurred in the aftermath of the first large-scale civilizations. They were built on top of neolithic agriculture which seems, by the way, to have been remarkably peaceful. A possible disadvantage is that women appear to have done all the work—maybe they still do. Anyway, on the basis of neolithic agriculture, man began to build up the great river valley systems which increased wealth, concentrated power, and plunged very soon into cycles of aggression and conflicts of leadership of the most appalling kind. In fact, the period I mentioned earlier—the four of five hundred years of civil war in China—seems to have reflected the horrendous power struggles arising in China as the river valley systems began to increase resources and centralize power.

But in the half millennium before the first Han Dynasty, all round our planet man witnessed the rise of the world's great ethical systems. Buddhist, Taoist, Confucian, the Hebrew prophets, the Platonists, the birth of Christian faith —these great new visions of man's moral reality grew up during these fateful centuries. There was, as it were, a conversion of people's minds to the realization that the way they were running their society—with violence, with rapacity, with greed, with self-indulgence, with leadership gone lunatic and crazy, with Assyrians coming down like wolves on the fold, with endless civil wars, with all the miseries of greed and power—could not represent the final reality about human nature and human society.

The arrogant idols of the tribe, the greedy idols of the market were seen to be insufficient, more than that, to be lethal and deathly. Instead came the vision of a god to be worshiped "in spirit and in truth." I believe we could be on the way to confronting a new moral reality on a similar scale today. And the vital and absorbing thing about this new moral reality is that it is not just based simply upon the insight of sages and prophets. It can now be based upon a most accurate and scientific vision of how this universe actually works. On the one hand, it is immensely powerful, pouring out on us the infinitely fierce and destructive energies of the sun's radiant energy. On the other hand, it is mediated to life through mechanisms so delicate that they must be treated with respect, with tenderness, with care.

In fact, if we are greedy about this delicate planet, we shall simply have no planet. We can all cheat on morals. We all know that and I suspect we all do it. We rise above exhortations. We can forget moral imperatives. But today the morals of respect and care and modesty come to us in a form that we cannot evade. We cannot cheat on DNA. We cannot get round photosynthesis. We cannot say I am not going to give a damn about phytoplankton. All these tiny mechanisms provide the preconditions of our planetary life. To say we do not care is to say in the most literal sense that "we choose death."

I would therefore put to you what I could call a modest hope. It is possible that in the latest age of turbulence and disaster, our science and our wisdom are coming together and our faith and our reality are beginning to coincide. And if Stockholm is a place where this process begins, let us all thank God that we were here when it started.

Unity Through Diversity

By RENÉ DUBOS

All human beings have the same biological needs, but they have developed an immense variety of social and individual demands. These demands differ so much from one nation or one ethnic group to the other that no environment can be equally desirable for all human beings, even if their fundamental needs could be completely satisfied.

The paradox inherent in the dual nature of man—namely the biological uniformity of mankind and the social diversity of human life—is at the heart of the questions to be discussed by the United Nations Conference on the Human Environment. A global approach is essential for dealing with the ecological and economic problems of the spaceship earth which affect all of us, but each human settlement has problems of its own which require local solutions.

33

Biological uniformity is a unifying force in the world, whereas social diversity is at the source of many conflicts. Yet I believe and shall attempt to show that social diversity, despite its disruptive effects, can generate a new kind of global unity more creative than that resulting from biological uniformity.

The various human races have a common origin, far back in the Old Stone Age. Although they have developed minor genetic differences as a consequence of different ways of life under different environmental conditions, all members of the human species still are essentially identical in anatomical structure, physiological attributes, and psychological drives.

Man continues to change, of course, as he adapts to new ways of life in new environments, but his adaptations are social and technological, rather than biological. Wherever he goes and whatever he does, he can survive and function only by maintaining around himself a micro-environment which resembles very closely the one under which the human species acquired its biological identity. Man can move today in outer space or on the bottom of oceans but only by remaining within enclosures that provide him with conditions similar to the terrestrial atmosphere—as if he had to be permanently linked to mother earth by an umbilical cord.

In a purely biological sense, it is therefore possible to define with precision an environment biologically suitable for human life. But actual needs are much more complex than biological needs because they reflect historical influences, contemporary social forces, and aspirations for the future—all factors which are peculiar to each human group. What people need is so largely determined by the traditions and expectations they inherit from their respec-

34

tive cultures that the phrase "the essential needs of man" has only limited usefulness.

Even the satisfaction of essential food requirements is profoundly affected by cultural factors. Biologists evaluate nutrition in terms of proteins, carbohydrates, fats, vitamins, and minerals; but a particular article of food can be selected or rejected for positive or negative symbolic reasons purely derived from social history. Americans are even more reluctant to eat horsemeat than Frenchmen are to eat cornbread.

Thus, while the biological qualities of a given environment can be defined in scientific terms applicable to all human beings, its social and economic qualities differ from one group to another, depending upon past experiences and value judgments.

Irrespective of cultural considerations, man's future is linked to his wisdom in using the resources of the earth. The total reserves of mineral ores and of fossil fuels may be sufficient to keep industry going at its present rate for several decades. New technologies, furthermore, will probably soon make it possible to produce large amounts of electric power and a variety of substitutes for the natural resources which are becoming scarce. But while there is no end in sight for human inventiveness, there are limits to the extent of human intervention that the earth can safely tolerate. Any further qualitative or quantitative change in technology must be weighed on the sensitive balance of global ecology.

Environmental degradation inevitably causes a degradation of human life, even when it does not interfere with fundamental biological activities. For example, a shortage of potable water is not likely in the near future, but good-tasting water is becoming scarce; chemical pollutants and unnatural stimuli rarely achieve levels which are acutely dangerous, yet almost any level of pollutants can damage health on prolonged exposure. In brief, environmental degradation may not be the apocalypse that will destroy the ecosphere but it may soon reach a level dangerous for the quality of human life.

The ecological constraints on economic development have acquired urgency from the recognition that most of the surface of the earth will soon be completely occupied and utilized. The colonization of the earth began, of course, long before the days of modern technology. But the difference is that now, men really occupy all land areas except those which are too cold, too hot, too dry, too wet, too inaccessible, or at too high an altitude for prolonged human habitation.

According to the Food and Agricultural Organization (FAO), practically all the most suitable lands are already farmed; future agricultural developments are more likely to result from intensification of management than from expansion into marginal lands. There probably will be some increase in forest utilization, but otherwise land use will soon be stabilized. In fact, expansion into new lands has already come to an end in most developed countries and is likely to be completed within a very few decades in the rest of the world. A recent FAO report states the probable date in 1985.

The Stockholm Conference therefore comes at a critical time in man's history. Now that the whole earth has been explored and occupied, the new problem is to manage its resources. The management of the earth, furthermore, must take the future into consideration. Most persons are concerned not only with their present needs but even more with the welfare of their descendants, and also with the fate of mankind as a whole. The phrase "stewardship of

36

the earth" denotes management beyond the here and now, for the sake of future generations. It implies also a close interdependence of agricultural and industrial activities. Modern agriculture, for example, can be efficient only where energy is available at low cost.

Agricultural efficiency cannot be measured only in terms of yields in plant and animal products. Another criterion is the relations of yields to energy used. And when modern agriculture is evaluated in this light, it proves to be rather inefficient. There are situations indeed where modern scientific farming recovers less energy in the form of food and other goods than it spends in its technological operations. The energy expenditure of modern farming includes the gasoline used for powering agricultural equipment, the electricity used for producing fertilizers, pesticides, tractors, trucks, and all other kinds of machinery, and also the energy used for irrigation or drainage.

Modern civilization would be inconceivable, of course, if the energy required by scientific agriculture had to come from human muscles instead of gasoline and electricity. But on the other hand, many of the modern agricultural practices are economically possible only where industrial energy is available at low cost. Like other aspects of human life, the future of land management and of agricultural production is consequently linked to the development of new sources of energy. In the final analysis, the stewardship of the earth thus involves the same kind of rational approach to the energy problem in agriculture as in industry.

Since human life is future-oriented, a desirable ecosystem is one which is capable of evolving in a direction favorable to the further social evolution of man. This anthropocentric view of environmental problems was almost unanimously held by the international experts who acted as consultants for the preparation of the report *Only One Earth.* Starting from the same set of scientific facts, these experts differed in their conclusions with regard to such controversial topics as the use of nuclear energy or pesticides, the desirability of further industrialization, or the optimum size of human settlements. But their differences of opinion did not originate from differences in knowledge of facts. The difficulties had their roots in the value judgments they put on these facts. Priorities differ from one place to another because, ever since man has abandoned the hunter-gatherer way of life, he has prospered only where he has extensively and profoundly transformed the surface of the earth.

Man has progressively transformed the wilderness into humanized nature, by clearing forests, draining marshes, irrigating deserts, extracting the minerals and fuels stored in the earth, and using them to modify natural environments or to create artificial ecosystems. Much of what we call nature is really man-made. Wherever man lives, the wilderness has been replaced by new ecosystems which have become so familiar that they are assumed to be of natural origin.

The transformations of wilderness have not always been successful; history is replete with ecological disasters. But the more significant and hopeful fact is that every continent can boast of man-made lands which have remained fertile or otherwise useful for immense periods of time. From Japan to Italy, from Java to Sweden, civilizations are built on man-made lands. Global history demonstrates that humanized nature can be ecologically sound, economically profitable, esthetically rewarding, favorable in biological and mental health—and can thus provide a substratum for the continued growth of civilization.

38

The interplay between man and environment commonly results in the successful transformation of a given ecosystem into a variety of human situations, each with its own characteristics. Very different cultures can thus coexist under the same set of natural conditions.

For example, the agricultural techniques, land policies, and behavioral patterns in the various islands of the South Pacific are determined not only by geologic and climatic factors but also by the skills and cultures of the early settlers—Polynesians, Melanesians, or Indonesians—and then later by the intervention of Western and Oriental people. Cultural attitudes more than natural conditions are responsible for the profound differences between Fiji, Tahiti, and Hawaii. These islands were initially settled by different groups of people and they display today the more recent influence of respectively English, French, or American colonizers.

The shaping of nature by culture can be recognized in many other parts of the world. For example, there is much climatic and geological similarity between England and northern France, yet the landscapes in these two countries have become different because they have been influenced by contrasting social and political histories. And similar remarks could be made concerning the differences between the Seattle region in the United States and Vancouver Island in British Columbia.

All over the earth, primeval forests have been transformed into human settlements which derive their characteristics—the genius of the place—in part, of course, from geology and climate, but even more from human activities.

The interplay between man and nature often corresponds to a true symbiosis, namely a biological relationship which alters the two components of the system in a

39

way beneficial to both. Transformation through symbiosis is one of the mechanisms which increases diversity among living things and ecosystems.

In many cases, scientific technology increases the range of the interplay between man and nature, thereby bringing out potentialities of them which would have remained unexpressed in the state of wilderness. Great bridges and highways are not only means of communication; they help to shape the human mind by revealing to the traveler details of the landscape and sweeping vistas of its organization that would otherwise remain unseen.

The proper use of technology can furthermore enrich the human significance of the earth's resources. A culture is conditioned by the resources available to it, but on the other hand, resources have meaning only in relation to certain types of culture. Termites are not a resource in technological countries, but they are used as food in certain parts of Africa. Iron ores are not resources for food-gathering people, nor was coal a resource in Europe until the nineteenth century. And uranium was not a resource for anyone until the development of nuclear science. In brief, a given material does not become a resource until a culture learns how to use it for a particular purpose. But once a material has become a resource it influences the subsequent evolution of that culture.

Scientific technology continuously develops new substitutes for conventional products and techniques; for example, aluminum and plastics are increasingly replacing iron derivatives; nuclear energy and perhaps solar energy can be used instead of wood or fossil fuels. Substitution, furthermore, usually means more than replacement; it can also broaden the range of experiences. Wood may no longer be needed for the production of heat; but a wood fire on the hearth generates sensual delights and unites

40

the family or the group. The horse cannot compete with the tractor in large-scale agriculture, but leisurely horseback riding puts the rider in contact with the rhythms of the living world.

Differentiation is an inevitable consequence of evolution, and symbiotic relationships enormously facilitate adaptation. Both differentiation and symbiosis increase diversity. From the international point of view, the trend to diversification may appear objectionable because it goes counter to the global unity which is symbolized by the phrase "One World." But on the other hand, the increase in diversity encourages the development of organs for communication and control; organs which generate a new kind of unity. Each region tends to achieve uniqueness as a result of the symbiotic interplay between the traditional culture of its people and the resources of its environment. This process of differentiation will continue despite the spread of international technology and it can be expected to generate uniqueness in certain types of industrial and economic activities.

In all living systems functional specialization eventually elicits the development of integrating mechanisms. The various parts of the human body are highly specialized but they function as a unity nevertheless through the operation of neural systems and hormones. An analogous process of integration is beginning to develop in the international order. From the simple personal contacts and barter systems which prevailed in the past, we have moved into the era of complex communications and exchanges operating over vast areas. And the process of integration will eventually involve the whole global body.

The threats of war, of disease, and of pollution, as well as the depletion of natural resources, are among the problems which first elicited integrating responses from the

41

international community. The Stockholm Conference is one of these responses. But in my opinion we must go beyond the development of mechanisms designed to prevent or minimize environmental defects. As the various nations and regions continue to differentiate—in economic activities and in life styles—they must elaborate new ways of relating to each other so as to become progressively integrated into organic wholes. Differentiation must always be followed by integration. To symbolize the need of achieving unity through diversity, the United Nations might eventually come to be known as the Integrated Nations.

In practice, a global approach is needed when dealing with the problems of the spaceship earth which affect all of mankind. But local solutions, inevitably conditioned by local interests, are required for the problems peculiar to each human settlement.

These two contrasting attitudes concerning the environment are not incompatible; in fact they complement each other. The national loyalty that we must develop toward the planet as a whole need not interfere with the emotional attachment to our prized diversity. As we enter the global phase of social evolution, it becomes obvious that each one of us has two countries—his own and the planet earth. We cannot feel at home on earth if we do not continue to love and cultivate our own garden. And conversely, we can hardly feel comfortable in our garden if we do not care for the planet earth as our collective home.

How Vulnerable Is the Ocean?
Introduction by Letitia E. Obeng

Man owes to the seas a debt he can never repay. For, with the oceans came the vital elements and supporting environments within which primeval life took form and the ancestry of man was cradled. Through phases, over eons of time, with unbounded benevolence, the oceans molded life from very humble Archeozoic beginnings. Cambrian environments within the seas permitted crude preparation for the adventurous Silurian leap from water to land and imprinted indelible animal marks in the sandstones of the earth. Without the sea and the ancient plants and animals it nursed within its bosom in those periods long ago, the planet earth might never have seen man in the Cenozoic era.

The course of events which took place within the granite and basalt depths of the oceans is still a mystery eluding the fertile brain and the deductive powers of man. Within this watery medium carrying sulphur, nitrogen, phosphorus, potassium, calcium, sodium and other salts, numerous levels of creation were traversed in time and space in the depths of the seas but, with calm majesty, the oceans have continued unperturbed their life-productive role.

The solid regions of the earth, perhaps only transient in cosmic time, occupy a mere third of its surface. The seas hold sway over land and decree its climatic conditions. They distribute heat and cold through currents, to make fair share of the sun's uneven heating of the lands. They raise vapors in large volumes and control winds, rains, pressures, and humidity which affect man and his land.

The seas perhaps hold the highest hopes for continued life. Yet, what does man do to the seas? Not only does he grab with greed the creatures of the sea, he turns nature's cradle for life into a receptacle for garbage and filth. This is man whose life blood contains sodium, potassium, and calcium in almost the same proportions as they still exist in the environment of mother sea which encouraged his birth.

How Vulnerable Is
the Ocean?

By THOR HEYERDAHL

At least five thousand years ago man started to rebel against
the nature that had bred him, and successfully nourished
him for perhaps a million years or more. It has been five
thousand years of technological progress and a continued
series of victories for the human rebel, the only mutineer
among the descendants of nature. Nature has yielded, tree
by tree, acre by acre, species by species, river by river,
while man has triumphed. He has been able to advance
by using the brain and the hands nature had given him to
invent and use tools, and to create new materials. One
might say that on the seventh day, when God rested, man
took over as Creator. He began to redesign the world and
mold it to his own liking. Century after century he has

45

worked without a blueprint to build a man-made world, each inventor throwing in an idea, each mason thrusting in a stone wherever a hand could reach. Only in very recent years have we begun to wonder what we are building. So far it has been taken for granted that any step away from nature was progress for humanity. Quite recently it has become more and more apparent that some of the changes man is imposing on his original environment could be harmful to himself; in fact they could even lead to global disaster. New inventions and new products continue to pour into the market while representatives from nations all over the world are assembled here in Stockholm today to try jointly for the first time to sort out the problem. What *are* we building and what *are* we doing to our environment? Can this unorganized rebellion against nature, this global building without plan, continue at its rapidly increasing pace without the haphazard structure collapsing over our heads? Is there something we can do to safeguard our existence in the changing world? Are we on the verge of destroying something that we ourselves and our descendants cannot do without?

Indeed the problem is manifold since civilization is complex and the world is large. It is my privilege here to raise some questions pertaining to one aspect of the problem: What are we doing to the ocean? As we all know, the ocean covers 71 percent of the surface of our planet. Is it vulnerable? And if so, is the ocean dispensable?

Unless we stop to reflect for a moment, I think most of us have the impression that the ocean is nothing but a big hole in the ground filled with undrinkable salt water, an obstacle to pedestrians and drivers alike, an abyss separating nations. On second thought we think of it as a conveyer for ships, a source of food, a holiday playground and, if we remember our schoolday lessons, we realize that it acts as a filter receiving dirty river water and returning it to our fields through evaporation, clouds, and rain. It is probably not unreasonable to assume that most people regard the dominant space of the ocean on our planet as more of a disadvantage than an advantage. With less land covered by the ocean, there would be more fields to cultivate, more resources available, and more space for the growing population to expand. The fact, however, is that the proportion of land to ocean is either extremely well planned or a remarkably happy coincidence, since it is this composition which has made life possible, at least in the form we know it, just on this one planet.

Whether we accept the story of the creation in the Bible or the findings of modern science, we all agree that life on earth began in the ocean. Neither God nor nature was able to create man from lifeless volcanic rock. The long and complex evolution toward man began below the surface of the sea when solar energy transformed the gases and minerals from eroded rocks into protoplasm and the first living cells. We do not know precisely how it happened, but after a certain period of time the first single-celled organisms were alive in the sea: the ancestors of all animals floated side by side with the ancestors of all plants. While many of them evolved into larger species, others continued to survive much in their original size and form until the present day, and are part of the marine micro-organisms we call plankton. The animal-plankton, termed zooplankton, sustained themselves by feeding on the plant-plankton, or phytoplankton, and the plant-plankton survived by feeding on dead and decomposed plankton of both kinds, and the minerals of the sea. In this process of metabolism the plant-plankton, and their descendants among the larger algae, began to produce oxygen in increasing quantities as they multiplied and filled the sunny surface layers of the oceans. So rich became this production of oxygen from the plant life in the sea, that it rose above the surface and into the earth's sterile atmosphere. When the atmosphere content of oxygen was high enough, some shore-stranded algae took hold, and after millions of years developed roots and leaves, becoming the first primitive terrestrial plants. But it was not until the content of oxygen in the atmosphere was approaching the percentage it maintains today that the first lung-breathing species were able to emerge from the sea. Animal life on the land and in the air had been impossible for hundreds of millions of years until plant life in the sea had produced sufficient oxygen. Then the animals of the sea developed into the creatures of the air and of the land. From then on, conditions were ready for the long evolution to the mammals, which terminated with the appearance of man. Man represents the crown of a mighty family tree with all its roots in the ocean. We cannot overlook this biological background.

We of the twentieth century, in spite of our supermarkets and superjets, have not become supermen who can cut off the unbilical cord to nature and survive alone as an independent species. We must never forget that we are part of a tremendously complicated system, part of a synthesis of biological species ranging from the oxygen-producing plankton of the sea, and forests of the land, to the food-producing soil and water, which can yield their indispensable supplies only through the activities of insects, worms, and bacteria, inconspicuous creatures which man in his ignorance usually disregards and even despises. Not even the invisible Creator, call him God or call him the Force of Evolution, could place man on the surface of the earth until all other biological species were already

48

there to support him. Only in recent years have we begun to understand the interaction of the extremely complex ecosystem, where every single biological species has its function in creating living conditions, birth and continued survival for other species higher up the ladder of evolution. On the very top of this ladder, man is balancing, ready to fall if for some reason he should lose the support of the species that preceded him and paved his road to existence. More than half of the biological pyramid supporting man at its apex is composed of creatures living in the sea. Remove them and the pyramid collapses; there will be no foundation to maintain life on land or in the air. We are all aware of the fact that an important part of the human food supply comes directly from the sea. In fact, modern economists are counting on a vastly increased output by ocean fisheries if we are to solve the growing problem of feeding the undernourished world of tomorrow. If we kill the plankton, we lose the fish, and thus drastically reduce the protein available for human sustenance. Man can limp along with very little in his stomach, but he has to fill his lungs. It takes weeks to die of starvation, days to die of thirst, and seconds to suffocate. If we kill the plankton we reduce to less than half the supply of oxygen available to men and beasts, and this at a time when forests are becoming scarcer than ever before. We, like all breathing species, will in fact be increasingly dependent on the plants in the ocean, since green landscapes rapidly recede before the spread of urbanization, industry, and the onslaught of modern farming, while asphalt, concrete, and barren sand dunes advance on previously fertile land. Since life on land is so utterly dependent on life in the sea, we can safely deduce that a dead sea means a dead planet.

49

Let us humbly admit it: If man is to survive, the ocean is not dispensable. If it is not dispensable, a pressing question naturally emerges: Is the ocean vulnerable?

If we ask the man in the street, if we ask the decision-makers, if we ask ourselves, we are likely to hear the answer: the ocean is not vulnerable; it is too vast to be damaged by little man's activities; it is an enormous self-purifying filter which has taken care of itself for millions of years and will continue to do so forever. We are all familiar with the phrase: "a drop in the ocean."

Since the morning of time, thanks to gravity, nothing has dropped off the earth, into space. Nothing, except a few truckloads of gear recently transferred to the moon. Millions of years with natural pollution. Millions of years without human industry, when nature itself was a giant workshop experimenting, inventing, producing, and throwing away waste. Waste by incalculable billions and ever more billions of tons of rotting trees, dead flesh, bones, excrement. Whether we measure in weight or in volume, the wastes of all the world's industry amount to nothing during the few decades of our technical era, as compared with the hundreds of millions of years of volcanic eruption, global erosion, and untold generations with death and decay. Man is not the first manufacturer; why should he become the first polluter? Our spaceship is built with an ocean to take care of pollution. So why worry?

Admittedly, man has started to manipulate atoms, both taking them apart and grouping them together in different manners. But so did nature before him. Nature's own workshop invented and produced neutrons and electrons, radiation and gravity. Gases, liquids, metals, and living cells. Vitamins and chromosomes. The beating heart and the thinking brain. The photographic eye and the listening ear. Nature combined molecules into darting shrimps

50

and lumbering elephants. It turned fish into birds, and beasts into man. Nature found a way to make oranges out of soil and sunshine. It invented "radar" and installed it into bats and whales; it produced shortwave transceivers for beetles and butterflies. The principle of jet propulsion was successfully tested out on squids, ages before nature created the human brain. The human brain is itself the most complicated computer of all times, packed in a bump-proof case, with a body to carry it around at its own will. As we have seen, it marked the peak of earthly invention, the latest model delivered from the workshop of spaceship earth.

Where are the endless masses of waste discarded along the production line? Gone. But not into space. Transformed. In fact, transformed into new living matter. The life cycle, the ecosystem of spaceship earth, is the closest one can ever get to the invention about all inventions: *perpetuum mobile.*

When nature could produce and pollute, why cannot little man? Volcanoes and sandstorms have sent fumes and dust into the air as long as life has existed, while dead fish and plankton have rained toward the bottom of the sea together with the silt from fields and rivers. There has never been any ultimate pollution of the land or of the air: it is all being flushed into the ocean. Of course, the land has been littered with dead leaves, fallen trees, rotting corpses, fermenting dung. But the rain is there to clean the air and to wash the rocks. Bacteria are there to transform death and subsequent decay into life and subsequent lovemaking. And gravity is there to drag the decomposed surplus by way of nature's drains into streams and rivers bound for the earth's ultimate sink: the ocean. Only the ocean has no outlet for solid waste. But silt and filth from the land are received as magic fuel for the billions of tons of plankton which helps to keep the ocean fresh and clean.

So why the fuss about little man's pollutants? Why can't *we* pollute when nature can? Simply because the pollutants of modern man have suddenly become basically different from those of nature.

Like nature, we have started to manipulate molecules. We have started to understand some of the most ingenious inventions that preceded our own existence. But when we take molecules apart and put them together in our own way, we get, to our surprise and delight, new and amazing substances never invented by nature although they appear to be highly useful to man: plastics, D.D.T. and an endless series of other effective insecticides, detergents, and chemical products hitherto unavailable to planet earth. Why had nature overlooked the usefulness of these inventions? A sheer oversight? Today we can really get our laundry white and can exterminate every trace of insect life. We spray leaves and trees, fields and swamps. We spray bugs and worms, bees and butterflies. We spray into the air and down on the soil. We wash and we spray and flush the detergents down the drain. We mass-produce. Industrial plants grow like temples around cities and lakes, along roads and rivers. Toxic wastes seep into every stream, pour into every sewer. Rain and drainage and man-made sewers carry pollution away into the sea. The mighty currents of the ocean churn around and carry away all poisonous waste. Carry away? No. Carry around. The ocean is as round as the earth, landlocked in every direction, with thousands of inlets, but not a single outlet.

Why cannot the oceanic filtering system function for technological man as smoothly as it functioned in the recent decades of our parents, or in the days of ape men, monkeys, or dinosaurs? Cannot all refuse, now as before, be re-used as fuel for the biological machinery? No. And this is precisely where the environmental problem has begun. Man has recently created materials which nature wisely avoided because they had no place among the perfectly interlocking cogwheels of the global ecosystem. Through technology, man has begun to throw extra bolts and nuts into a ready-made and smoothly operating machinery. Although it has only begun with our own generation, year by year, day by day, man's production of nontransformable, nondegradable materials has increased in a virtually precipitous curve. Waste and refuse are littered everywhere and we try to sweep it all into the ocean as if under a carpet. We realize that an increasing quantity of our modern waste is both toxic and nondegradable. But it doesn't matter; the ocean is endless. The ocean is deep.

This is, in fact, the second misconception. For the ocean is not endless and its depth is greatly misleading.

Just as we have seen that the age of the ocean is no guarantee of its invulnerability in technological times, in the same way we shall see that its size is no better guarantee.

Anyone who has drifted from one continent to another on a bundle of reeds or a log raft cannot help being struck by the fact that the ocean is just another big lake. It is enough to place ten Lake Eries end to end and they will span the Atlantic from Africa to America.

With technical progress, distances have dwindled, dimensions have changed. With the astronauts we begin to see that the ocean has limits like any lake; we begin to see our planet as a lonely spaceship. A spaceship without exhaust pipe. We have begun to realize that no chimney is tall enough to pierce the atmosphere and send our fumes into space, nor is any sewer long enough to pipe our pollutants beyond the borders of our common sea. For primitive man standing on the beach the blue ocean ran into the blue sky; for us it curves beyond sight and falls back on us from behind. Sea and soil, fumes and sewage, all are here to ride with us forever in some form or another on the thin crust of our spinning sphere. Even so the sea is large compared to man. After all its average depth is 1,500 meters (or about 5,000 feet) running, in places, down to more than 10,000 meters (or 30,000 feet). If this average depth of 1,500 meters were stretched out along a road, a runner could cover the distance in less than four minutes. And, if we were to represent the world ocean on a topographic globe, it would be impossible to smear on a coat of blue paint thin enough to represent the depth of the ocean in true proportion.

Although the ocean layer, in a global perspective, is thus much thinner than people suppose, only a fraction of this water needs to be polluted to kill all life. Marine life is concentrated near the surface and, again, most of the surface life is concentrated near the shores. Why is only the upper limit of real importance to marine life? Because marine life, directly or indirectly, depends on the plant-plankton as basic food; and plant-plankton can live only where the sunlight can penetrate the water in sufficient quantity to permit photosynthesis. This upper section, in the sunny tropics, is only 80 to 100 meters deep, whereas in the northern latitudes, on a bright summer's day, it has a depth of only 15 to 20 meters. Below this thin layer, life is immediately very restricted since it depends on the sinking of decomposing plant and animal remains from the sunlit waters above.

Why is most of this already restricted surface life further concentrated near the shores? Because marine plant life, in addition to sunlight, needs mineral nutriments. In the

54

coastal areas some of these are brought down by rivers and others are returned from the bottom of the sea. The ocean bottom is rich in these nutriments which have rained down from decomposing organisms near the surface and they can only return to the surface in areas where strong underwater currents and upwellings occur, such as near continental shores. An estimated 90 percent of all marine life is found in shallow coastal areas generally referred to as the continental shelves. These important areas represent only 8 percent of the total surface of the oceans, and, of course, only a fraction of 1 percent of the total ocean volume. When we speak of farmable land, we refer only to the usable surface soil; we never count the volume of sterile rock beneath as deep as it may go. Man harvests the sea as he harvests the land; let us therefore not fool ourselves by relying on the depth of the ocean any more than we rely on the depth of the land. It is, in fact, the very areas where life is concentrated that man pollutes most.

Most pollutants come from the land. How many rivers today have drinkable water at their mouths? In fact today some rivers would not flow at all without factory and urban effluents. All the polluted rivers and all the sewers of the world empty their toxic refuse directly onto the continental shelves. This is also the nearest and cheapest dumping area for the enormous quantities of unwanted chemical poisons shipped away from the shore. What is so dangerous that we are afraid of storing it on land, we dump overboard guided by the principle "out of sight, out of mind." As an example, incalculable quantities of poison, including entire shiploads, have both openly and secretly been transported from industrial countries in Europe in recent years, to be dumped in the presumably bottomless North Sea. This sea is so shallow that at certain points, the depth is only thirteen meters, and during the Bronze Age, half the North Sea was dry land.

Much has been written for and against the American act, repeated by others, of dumping vast quantities of nuclear waste and obsolete war gases in the Atlantic with the excuse that it was all sealed in special containers. There are already enough examples in shallower waters like the Irish Sea, the English Channel and the North Sea of similar foolproof containers moving about with bottom currents and cracking open with the result that millions of fish are killed or mutilated. In the shallow Baltic Sea, seven thousand tons of arsenic were dumped in cement containers forty years ago. These containers have now started to leak. Their combined contents are three times more than needed to kill the entire population of the earth today. It may be difficult to find means of stopping ocean pollution from seepage and sewers, but deliberate dumping in the sea of material too dangerous to keep in sight should be considered a criminal act. It could easily be forbidden and the offenders should be most severely punished under international law. In fact there is no such thing as "national waters." The ocean is in constant movement. Only the solid ocean bottom can be mapped and divided between nations, not the mobile water above it. If you launch a raft off the coast of Peru, it will be carried with the currents to Polynesia. If you set a reed boat afloat off the coast of Morocco it is carried to Tropic America. This illustrates that there is no such thing as territorial waters for more than days at a time. What are territorial waters of Peru today are territorial waters of French Oceania tomorrow. What are territorial waters of Morocco now will shortly become territorial waters of various states around the Caribbean Sea.

Even though it is extremely important to put an immediate end to deliberate ocean dumping, this is only part of the problem. By far the greatest quantity of toxic refuse constantly reaches the sea in a much less spectacular way from agricultural fields and urban and industrial sewers.

To visualize the immense quantities of solid and dissolved waste and fluid chemicals of all kinds which every minute flow into the ocean from the shores of all continents, we should imagine the ocean without water, as a big, empty depression. The fact that every river in the world empties into the sea without causing an overflow, makes us subconsciously think of the ocean as a witch's cauldron whose contents never come over the brim, no matter how much is poured in. We are apt to forget that the ocean has its own sort of outlet: the evaporation from its surface which permits only pure water to escape while all our poisons, all our solid and dissolved wastes, are left to accumulate forever in the pot. Visualize, then, the ocean as a dry and empty valley ready to receive only what man is pouring in. The rising level of toxic matter would clearly be visible from all sides. A few examples picked at random will illustrate the input we would witness.

French rivers carry 18,000 million cubic meters of liquid pollution annually into the sea; the city of Paris alone discharges almost 1,200,000 cubic meters of untreated effluent into the Seine every day.

The volume of liquid waste in the Federal Republic of Germany is estimated at over 9,000 million cubic meters per year, or 25.4 million cubic meters per day, not counting cooling water which amounts to 33.6 million cubic meters per day. Into the Rhine alone, 50,000 tons of waste are discharged daily including 30,000 tons of sodium chloride from industrial plants.

A report of the United Nations Economic and Social

Council states that we have already dumped an estimated billion pounds of D.D.T. into our environment and are adding an estimated hundred million per year. Most of this ultimately finds its way to the ocean, blown away by wind or washed down by rain. The total world production of pesticides is estimated at over 1,300 million pounds annually. The United States alone exports over 400 million pounds per year.

Less conspicuous than the constant flow of poisons from the shores is that even the tallest chimneys in the world send their pollutants into the ocean. The densest city smog and the darkest industrial smoke will slowly be carried away by the wind only to descend with rain and snow into the ocean. Cities and industries are expanding day by day, and so far, in America alone, waste products in the form of smoke and noxious fumes amount to a total of 390,000 tons of pollutants every day, or 142 million tons every year.

The ocean is the ultimate sink for all pollution disposed of in modern communities—even what we try to send up in smoke.

It is common to think of ocean pollution as waste thrown overboard from thousands of ships, or a spectacular accident with an oil drill springing a leak, such as in the Santa Barbara accident, or a super-tanker running on a reef, such as the *Torrey Canyon* in the English Channel. It is not the spectacular accidents that hit the headlines that should scare us most but the daily, intentional, almost inconspicuous discharge of crude oil from navy and merchant ships all over the world, above all the routine cleaning of oil tankers.

The whole world was upset when the *Torrey Canyon* unintentionally spilled 100,000 tons of oil in the English Channel. Every year more than 100,000 tons are intentionally pumped into the Mediterranean, an almost land-locked sea. The traffic of oil tankers in the Mediterranean is increasing at a tremendous rate and, according to a report made at the University of Trieste, it is estimated that in the year 1980 500 million tons of crude oil will be unloaded in Mediterranean ports. A recent study showed that for every square kilometer in the Mediterranean south of Italy there are 500 liters of solidified oil. In recent years visible pollution has begun to appear even in the largest oceans. In 1947, when the balsa raft *Kon-Tiki* crossed 4,300 miles, or nearly 8,000 kilometers of the Pacific in 101 days, we on board saw no trace of man until we spotted an old wreck of a sailing ship on the coral reef where we landed. The ocean was clean and crystal clear. In 1969 it was therefore a blow to us on board the raftship *Ra* to observe from our papyrus bundles that entire stretches of the Atlantic Ocean were polluted. We drifted slowly past plastic containers, nylon, empty bottles and cans. Yet, most conspicuous of all was the oil. First off the African coast, next in mid-ocean, and finally in front of the Caribbean islands, we drifted for days on end through water reminding us more of a city harbor than of the open sea. The surface, and as far as we could see through the waves, was littered with small clots of solidified black oil, ranging in size from that of a pin-head or a pea to that of large potato. When we repeated the same general itinerary with *Ra II* the following year, in 1970, we carried out a day by day survey and found sporadic oil clots floating by within reach of our dipnet during forty-three days out of the fifty-seven days the crossing lasted. Forty-three days out of fifty-seven, and in this time we had covered 3,270 miles, or 6,000 kilometers, of open sea from Safi in Morocco to the Caribbean Island of Barbados. With our sail, we moved faster than the floating pollution and since the equatorial current runs westward with an average speed of half a knot into the birthplace of the Gulf Stream, the flotsam we passed en route from Africa to America in 1969 was, during our second crossing a year later, washed ashore in tropical America, or else displaced more than 4,000 miles and thus drifting on its way back toward Europe across the North Atlantic. A detailed report on these observations was at the time sent to the permanent Norwegian delegation to the United Nations, and has been published by the Secretary General as annex to his report to the Stockholm Conference: *The Sea: Prevention and Control of Marine Pollution.* A laboratory analysis of the various samples collected shows a wide range in the level of contents of nickel and vanadium, revealing that they originate from different geographical localities, which again proves that they do not represent the homogeneous spill from one leaking drill or one shipwrecked tanker, but the accumulating waste from the routine washing of the world fleet of oil tankers.

The problem of oil pollution is a complex one. Various types of crude oil are toxic in different degrees, but they all have one property in common: they absorb other chemicals, notably insecticides, like blotting paper. Thus on the surface of the ocean, D.D.T. and other chlorinated hydrocarbons which do not dissolve in water and do not sink are attracted to oil slicks and oil clots, where in some cases they can be concentrated to an even higher percentage than when originally bottled with dissolvents for spraying purposes.

From the *Ra* we observed that larger oil clots were commonly covered with barnacles, marine worms, and small crabs which did not feed on the tar lumps but used them to hitch-hike across the ocean. However, these riders are attractive bait for fish which swallow the whole cluster.

These ever-present oil clots can hardly, even if they are not baited with barnacles, avoid getting into the bodies, gills, and baleen of filter-feeding fish and whales. Their effect naturally depends on their concentration of inherent and acquired toxic matter. In the famous Sargasso Sea, the oil lumps are now so common that recently an expedition of marine biologists had to give up working with dragnets on the surface, since the mesh of their nets was constantly getting plugged up by solidified oil, and their catches literally brought in more oil clots than seaweed.

What happens to all this floating oil? The optimist says that very likely the lumps may gradually disintegrate and sink to the bottom. Perhaps this is true, but certainly not before they have done harm to surface life and the coastal flora and fauna. The tourist industry is concerned about the oil washed up on the beaches, but it is time to get more concerned about the oil washed against the inaccessible rocks. Maybe the oil clots gradually sink, but so far they are certainly only accumulating, disclosing their presence and leaving their effects in an ever more conspicuous way. Fishermen have already begun to notice a grayish belt, darkening from year to year, just above water level on coastal cliffs and boulders. Particularly in the eastern and southern Mediterranean and on several of the local islands, the yellowish rock is becoming discolored from yellow to gray and even black in a belt six to eight feet wide. In many areas this belt along the waterline, once covered by seaweeds, clams, mussels, limpets, and all sorts of crustaceans and small fish, is now dead. Completely sterile. Nothing undulates or darts about. The once-yellow rock seems like coke, and in many places sizable clots of oil are hammered into the pores like sheets of tarmac. And this is happening in what is the cradle for the bulk of marine life, since most species have to pass one stage of their life cycle on the rocky shores of islands and continents.

We hardly need to be reminded of the fact that even a drop of liquid oil will spread out over a large surface area of water, and the thicker the film of oil, the less is the photosynthesis and the production of oxygen.

In spite of all this, it is not the floating oil clots we should fear the most. Their presence shows us that our planet is not endless and that visible human waste is beginning to bridge world oceans. The oil clots in the sea tell a similar story to that of the refuse on the beach. Visible pollution is seen by some as a symptom of welfare. If we want to bless its presence, however, it should be as an eye-opener. Each empty plastic bottle or tube, each can, each oil clot we see along the roadside or on the beach should ring a bell and make us think of invisible pollutants already lost in the soil or the sea: liquids and particles not discernable to the naked eye. Man does not dispose of empty packing only. It is the lost contents and not the empty containers that should worry us. Where is the spray, the paste, the powder, the liquid no longer inside the empty packing? D.D.T. is now found in antarctic penguins and arctic polar bears living far from any area where insects have been sprayed. Even in the blubber of each of twenty whales recently caught by special licence for testing in the arctic current off the Greenland coast, six insecticides, including D.D.T., were present. These whales, born and raised off the East Greenland glaciers, had never been near the shores of agricultural lands. But the ocean is revolving and with it the pelagic plankton which is eaten by the krill that forms the basic diet of the whale. The plankton, just like the oil, has the property of absorbing, assimilating and concentrating the insecticides. These ever-present micro-organisms feed by filtering the sea water,

62

and retain the undesired chlorinated hydrocarbons together with the desired organic nutriment. What the plankton absorbs goes into the flesh of fish and finds its ways back to man again. Most of the plankton can hardly swim, and less so can the oil clots and toxic chemicals, but the ocean itself takes care of the transport: like the winds it is a constant conveyer. A good example is that a certain type of D.D.T. sprayed on crops in east Africa was found and identified months later in the Bay of Bengal, a good 4,000 miles away.

In the short period that the United Nations representatives are meeting here in Stockholm to discuss the problems of our environment, an estimated 50 million pounds of additional pesticides will have time to reach the ocean to join what has already accumulated there, not to mention the vast quantities of toxic chemicals pouring in from both industry and household.

I appeal to the multinational representatives in Stockholm to put aside all immediate personal and national interests and to be aware of the immense responsibility they have toward present and future generations.

Let us hope they bear in mind that the ocean currents circulate with no regard for political borderlines, and that nations can divide the land, but the revolving ocean, indispensable and yet vulnerable, will forever remain a common human heritage.

Economics of an
Improved Environment
Introduction by Saburo Okita

A central problem in the economics of an improved environment concerns the necessity of a slowdown of economic growth in order to prevent further deterioration of the environment. Whether or not a slowdown is necessary provokes sharp differences of opinion.

The first school of thought argues that growth is essential. It is essential in order to finance the investments necessary to prevent pollution and to improve the environment by a better allocation of resources.

The second school of thought, also pro-growth, stresses the great potentiality of science and technology.

Neither of these schools sees any need for fundamental changes in the nature and foundation of economic policy. The environmental issues, as they view them, are mainly matters of setting priorities in the allocation of resources.

However, more fundamental problems will arise if we pursue the third line of thought, which maintains that, in view of the finiteness of the globe's surface, the finiteness of natural resource deposits, and the overall life-support systems of the earth, the exponential growth of population and economic output will inevitably lead to a collapse of the growth mechanism itself. To avoid the catastrophic consequences, conscious efforts must be made to design policy and persuade people to slow down population and economic growth to a "zero growth condition" or "state of equilibrium."

Unfortunately, the conventional wisdom of economic policy is that it is desirable for industrially advanced countries to achieve more rapid economic growth, creating increased demand for imports and thereby stimulating the economic growth of underdeveloped countries.

If viable policies for environmental improvement are to be designed, if a global strategy of developed countries accelerating growth is to be accepted, then conventional approaches must be thoroughly reviewed.

Economics of an Improved Environment

By GUNNAR MYRDAL

I

The anxieties now expressed by biologists and other natural scientists working on the world's ecosystem concerning the impending depletion of irreplaceable natural resources such as water, energy, some crucial metals, and arable land, and the pollution of our environment—air, water, land, animals, indeed our own bodies—should rightly have important consequences for development planning in developed and underdeveloped countries and primarily for economic theory both on the macro and micro levels.

Of the two environmental factors, pollution is, on the whole, a quite recent worry. In regard to depletion, there have been several development experiences reaching back more than a century that have conditioned economists as well as the general public not to take such warnings about a deteriorating environment too seriously. And, indeed, it can now be witnessed how economists rather generally, with some exceptions, including the present speaker, tend to belong to the "optimists," who are inclined to discount these warnings as exaggerated.

One such early conditioning sequence of events was the refutation by experience of Malthus's theory and the implicit forecast that what he called the tendency of "geometrical" growth of population, if not checked by "moral restraints," would come to press on the agricultural resources of land. These Malthusian thoughts were accepted by Ricardo and the whole early school of classical economists. They provided the basis for Ricardo's harsh theory of wages and of distribution in general. Indeed, they led him to vaguely envisage as a result of development the stationary state, with rising land rates, falling profits, the ending of real capital accumulation, and wages becoming stabilized at the production costs of the labor force, all based on a balance of the forces of growing scarcity of land and the tendency toward population increase.

Quite contrary to this gloomy forecast—except indeed for the underdeveloped regions—the era from Malthus up to the present has been one of historically unprecedented economic growth and also gradually rising levels of living even in the lower-income brackets. This has happened in spite of periodic interruptions by depressions and even devastating wars. This growth generally accelerated after the Second World War.

In hindsight, it is possible to account for the major alterations that had this result: among them rapid technological advance in agriculture as well as industry; from

68

the middle of the nineteenth century, the huge imports of food to Britain and Western Europe generally from the emigrant settlements in the areas of the New World where the natives were not so numerous and could be killed off or segregated in various ways; and still later the spontaneous spread of birth control.

Another series of experiences conditioning us to "optimism" has been the following: through all my working life as an economist there have been fears about the pending scarcity of raw materials. Even the Atlantic Charter contained a passage expressing this fear and requesting equal access to raw materials. From time to time there have been large-scale studies giving similar warnings, among them the *Paley Report* in the fifties and more recently the report on *Resources and Man* by the National Academy of Sciences in the United States.

But in actuality there has always been a good supply of raw materials. Indeed, the development prospects for underdeveloped countries have continually been endangered by oversupply and the low prices of the primary products depending on land resources, that make up the bulk of their exports.

Again, it is possible *ex post* to explain why these newer fears have so far not been justified by later developments. A main cause has been the never-ending discoveries of new supplies of raw materials, particularly of oil and various metal ores. In addition, there have been technological inventions making it possible and economically profitable to create economies in the use of raw materials, and also to replace some of them by less scarce raw materials or by manufactured products produced from them.

Because we in the developed countries have been permitted to retain that important element of the faith in human progress prevailing in the era before the First

World War, we all have come to trust in continuous and boundless economic growth. In that respect there has been no difference between developed countries with different political systems. And after the hurricane of liberation from the colonial power system that after the Second World War swept over the globe, the underdeveloped countries raised demands for development, basing their planning on the same faith in the unlimited space for growth, could they only overcome the lack of capital and other obstacles and inhibitions for development.

II

We have so often heard the cries that the wolf is coming that we have become accustomed not to take them seriously. From my studies, though naturally inexpert, of what my colleagues in the natural sciences have more recently reported, I have, however, become convinced that we must finally recognize and prepare for the fact that there are limits to a growth whose component elements all follow an exponential curve.[1]

All estimates upon which the warnings for depletion and pollution are founded are utterly uncertain, as another of the lecturers in this series, Professor René Dubos, has authoritatively explained: "The existing knowledge of

---

[1] A main reason why I am inclined to take a more gloomy view of the future than most of my colleagues among the economists is that I foresee that this time an undirected, natural development will not save us. We will need to take planned government actions on a large scale to defend our environment. And I mistrust both the will and the capacity of governments to decide upon and effectuate such policy to the extent needed. To this I will come back, as I proceed.

70

the natural sciences is not sufficient to permit the development of effective action programs." Many of these estimates are still of a highly controversial nature. In particular, further discoveries and inventions can only very broadly be surmised. Future policies which might even come to steer and thereby influence the scope and direction of science and technology are in principle not possible to predict, as they will depend upon the way in which people choose to act and react. History is not a blind fate.

But these reservations against the inexcusably careless manner in which so-called futuristic research is now often pursued do not decrease the necessity of long-term planning. What happens in our children's time and even further ahead, if this type of exponential economic growth continues, does matter to us as citizens of our nations and of the world.

And as has been demonstrated in a broad way, the reservations I have piled up merely imply that the future, when uncontrolled growth comes up against serious limits, is somewhat indeterminate, but usually within the range of only one or at most a few generations.

In any case, with the unprecedentedly rapid and still accelerating growth of the world's population, which we must now take as fairly certain to continue for many decades—to this I will return—we shall invite catastrophic developments, unless we are prepared now to introduce and enforce various restraints and deflections of production and consumption and, indeed, our ways of life. If we in the developed countries were alone in the world, but still had free access to the primary products that underdeveloped countries are now exporting, the threatening consequences of uncontrolled growth might not disturb us too much for several decades ahead, provided that we took reasonable measures against pollution. The situation in the underdeveloped world is very much more serious and the impending dangers more immediate, which I shall revert to later.

III

As usual, the economists have in their work reflected and rationalized the common inclinations among the general public in the societies they are living in and servicing. At the same time they thereby support in a mighty way the ideology and psychology of continuous and unlimited economic growth.

There are, in particular, two traits in modern economic theory that reflect this accommodation to prevalent thinking and which, consequently, have to be given up, or altered in a radical fashion, in order to build up an "economics of an improved environment," to quote the title given me for this lecture.

One is the concept gross national product, of which national income is an affiliation, and the use made of this concept. It is no exaggeration to state that modern economic theory has more and more organized itself around this concept, by breaking up the GNP into its constituting elements and seeking their determinants and the causal interrelations in the economic system made up by them all.

Although I have no possibility to demonstrate it in this brief lecture, I must stress that GNP is a flimsy concept in developed countries and, for various reasons, even very much more in underdeveloped countries.[2] These defects

[2] Mydral, "A Contribution Towards a more Realistic Theory of Economic Growth and Development," in *Mondes En Dévelopment*, No. 3, 1972.

72

stand out as particularly relevant in regard to problems of a long-term character, as those I am discussing today.

Several of the many thousand categories of income and cost elements are thus defined in a grossly arbitrary way; some elements are not included at all in the calculation of the GNP. And any attempt to introduce ongoing depletion and pollution into the calculation of the GNP must fail, because of the gross uncertainties I have alluded to, and the consequent difficulty or even impossibility of quantifying these facts in the way needed for aggregation. They are therefore absent, and must remain so, from the calculation of the GNP.

Without being able to prove it today, I must add that attempts to define a broader concept of "social utility," including not only production but also all other social indicators, are doomed to remain on a level of useless speculation that even conceptually is not valid.

In addition to this comes not only the lack of reliable quantification of all the things that are deemed desirable or harmful, but also more fundamentally, the almost total absence of quantitative knowledge about "the coefficients of interrelations" between the various factors determining the movement of the social system as a whole. This becomes particularly apparent when, as we certainly should, we include attitudes, institutions, and political forces as they operate in the "circular causation with cumulative effects," which governs the movement of the social system.

In the very short-term economic analysis—and in developed countries though, as I firmly hold, not in underdeveloped countries—the GNP as now calculated may nevertheless have some indicative value. In all countries many individual figures from national accounts relating to ingredient elements may also be of use. But for the type of long-term problems, focused on the facts of depletion and pollution that are excluded from the calculations together with attitudes, institutions, and political forces, the GNP has to be thrown out as entirely inadequate to reality.

This is more serious, as the concept GNP is now being used as the kingpin of economic analysis of the development problems in both developed and underdeveloped countries. Indeed, this use of the GNP has become the expression and justification of the ideology and psychology of continued unlimited growth I have mentioned.

IV

At this point a few critical remarks should perhaps be added to the recently published so-called Rome Report, *The Limits to Growth: A Global Challenge*. It will probably have the useful effect of popularizing the ecologists' broad warnings of the necessity of giving up our expectations to continue on the road of unrestrained growth. But to the serious student it has grave defects in its very approach to the problems of both present trends and the possibilities and means of altering these trends.

To begin with, the Report uncritically accepts the concept GNP without any queries. Also for the rest, it builds upon, and aggregates in the most careless way, data that are extremely uncertain both in regard to economic growth itself and its various components. The data concerning threatening pollution and depletion are equally unreliable. Even a popular presentation should contain a reminder of this, particularly as it is of importance for the use of these data in a system analysis. The authors are, in other words, overselling their product in regard to the validity of the basic data.

Much more fundamental is the question of the realism of the Report's global "world system analysis." This analysis implies, to begin with, a non-consideration of the enormous and increasing differences and inequalities within countries and still more between countries (see below). To explain this, the Report states that "inequalities of distribution are defined as social problems" and then placed outside "the world simulation model," which only "calculates the maximum possible behavior of our world system," provided that there is "intelligent action on world problems, from a world-wide perspective."

A social scientist working on these very problems will be hard put to give any intelligible meaning to this assumption of perfect harmony in the world. Still less will he be able to outline how it could be brought about. Particularly in a pretended system analysis it is simply not possible to get away from the social problems merely by stating that they are not taken into account. The ecosystem has to be studied as part of the social system I have mentioned.

More specifically the Report places outside the "interactions" within the "world model" attitudes and institutions, indeed even the process of prices formation, while politics is only represented by stating a number of the results of abstract policy alternatives. Their system is, therefore, far from inclusive enough to have meaning.

The birth rate, for example, is quite rightly a factor, and a very important one, within their model. But it is certainly not a function only of the other factors within that model and the interrelations between them all. As we who have studied the demographic development in the several regions of the world know, the movements of these other factors are not even among the most important determinants of the birth rate. And the importance of them is not through the simple interrelations of the model. Indeed, those interrelations are fictitious.

Under these circumstances the use of mathematical

equations, and a huge computer, which registers the alternatives of abstractly conceived policies by a "world simulation model," may impress the innocent general public but has little, if any, scientific validity. That this "sort of model is actually a new tool for mankind" is unfortunately not true. It represents quasi-learnedness of a type that we have, for a long time, had too much of, not least in economics, when we try to deal with problems simply in "economic terms."

In the end, those conclusions from the Report's analysis which are sensible at all, are not different and definitely not more certain than they could have been reached without that elaborate apparatus by what Alfred Marshall called "hard simple thinking aware of the limitations of what we know."

V

The second trait in economic theory I referred to which implies an accommodation to the ideology and psychology of unlimited growth is that, fundamentally, economic analysis has retained its character of being carried out in terms of a theory of price formation in competitive markets. The preferences of all concerned are expressed in aggregated form by their demands and supplies. Production and resource allocation are steered by relative profitability.

With regard to depletion and pollution, a main fact is, however, the very heavy discounting of the future at the present time, represented by interest and profit rates. This implies that the time horizon becomes much narrower than should be accepted by our collective society, which must consider developments decades and centuries ahead. When, as has happened, economists argue that, as resources become scarce, the cost of these resources will rise so that depletion is avoided, they are not taking into account the fact that this reaction does not come early enough so as to be rational and sufficient in order to avoid depletion.

What is needed to realize the goals of society is a large-scale correction of the process of price formation: the conditioning of production, investment, scientific and technological research, and consumption to take a different course and, also in regard particularly to pollution, to get the whole population to behave differently in regard not only to what they demand for consumption but also, for instance, how they behave in the matter of wastage and refuse.

There are, in principle, two means of introducing and giving effect to policy measures against depletion and pollution. One is direct government regulation of people's behavior, forbidding them to do certain things and directing them to do others. The second way is that of price policies, inducing by means of charges and/or subsidies the price formation to give different signals to consumers and producers.

Now it can be said that economic theory is already well adapted for accounting for such alterations by policy measures of what is a "may" or a "must" in economic behavior of all participants in the process of price formation, and in addition what things actually come to cost and what income an activity renders.

Price formation is, of course, never thought to work within an abstract society but in an actual national community which has laid down its parameters. Economic theory is, for instance, not helpless in dealing with the effects of protectionist policy measures or of labor legislation, the latter ordinarily containing both types of interferences in the process of price formation.

The rule is, however, that the effects of these interferences have been rather marginal, leaving the process of price formation to be mainly ruled by spontaneous demands and supplies which operate within the established parameters. The new restraints and deflections, rationally motivated by considerations of long-term effects of the trends toward depletion and pollution but infringing upon people's impulses to follow their individual short-term preferences, must in comparison stand out as radical.

This opens first a serious political problem. How is it possible to move from a general awareness by the public of the dangers ahead, and of the policy-deciding instances acting on behalf of the public, to a preparedness to impose the controls needed? This would ideally imply a centrally imposed and enforced planning of almost all economic and, indeed, all human activity.

Implied is also an equally serious administrative problem: how can the controls be applied so that policy decisions really become effectuated, and now can that be done without necessitating a huge policing and controlling administration which would be financially too expensive and also too obnoxious to the people?

I will return to these two major problems, the political and the administrative, at the end of my lecture, where I will attempt to characterize present policy trends and make a short-term forecast of what will actually evolve in the way of realizing the ideals of the movement that have brought into being the present United Nations Conference on the Human Environment.

I would like to add here that from the point of view of the administrative problem there is all reason to operate as far as possible through price policies which raise much more easily manageable needs of controls, than through direct government interdictions and prescriptions. On this particular point, I believe I can speak for the whole profession of economists. But we would deceive ourselves if we believed that the goals could be reached without direct controls also, for instance in the vast field of waste disposal or in regard to international treaties directed toward the control of pollution transmitted through water and air.

I feel sure that economic science will increasingly be able to lay the foundation for such radical central planning which may be necessary to meet the pending dangers of depletion and pollution. It will imply, however, such alterations in our approaches that we would have the right to talk about a "new economics." To what extent such a development of our theory would influence government policy is another matter, however.

The concept of the GNP, and the whole structure of theoretical approaches built up with the GNP as a central axis, will have to be dethroned. And the process of price formation will be even further removed from being essentially concerned with aggregated individual preferences represented by demand and supply.

As economists as a rule are a rather conservative lot in so far as the main structure of their conceptions and preconceptions are concerned, it was not to be expected that they should have been in the vanguard for such a fundamental change of society and their own science. The main pressure will in the future years, as well as up till now, have to be exerted by those natural scientists who are studying the ecosystem.

VI

So far I have had in mind the developed Western countries, including also Japan and the two small Australasian developed countries. It could be expected that the problems would have taken a different shape in the East European Communist countries with their centralized and comprehensive planning.

We might first note that, at least until recently, their planning has been directed almost solely upon economic growth. In that respect they have not been different from the Western countries, even although they have defined the GNP somewhat differently. A change, implying increasing considerations of the problems of depletion and pollution, is now under way. These countries should more easily be able to move further in that direction as they have only to incorporate new goals for the central planning that already exists.

This does not mean, however, that they have escaped, or even can escape, the two serious problems, the political and the administrative ones, that I briefly characterized in regard to the Western countries. Indeed, when they have not been more successful in speeding up economic growth than they actually have been, the main reasons are, I believe, the necessity to take into account the inclinations of their consumers and, even more, the top-heavy, hierarchically structured bureaucracy which is needed for implementing their central planning.

In these countries prices are, in principle, not left free to be the result of a market process of price formation but are centrally decided upon as part of the planning. In regard to the determination of the prices, their economists have not been very successful in explaining and motivating it by a clear-cut theory. There have in all these countries intermittently been attempts to allow space for a somewhat freer play of demand and supply in markets, thereby decentralizing decision-making in the interest of efficiency and rational resource allocation.

I foresee that it is probable that the European Communist countries will retain their central planning, increasingly directing it not only on growth but also on improving environment, but, at the same time, giving price formation a somewhat freer play. Meanwhile, under the influence of the environmental crisis, the Western countries will be compelled to move toward more central planning, while retaining as much of a market economy as possible. This will tend to decrease the distance between the two types of economies. And there are other developments working in the same direction.

Fundamentally, the threatening advance of depletion and pollution gives rise to very similar problems, and even the needed policy adjustments and the administrative difficulties are broadly similar.

VII

In the global perspective, it is the great majority of people living in the underdeveloped countries that is most seriously and most immediately threatened by the environmental crisis. There are, among other things, to which I will come back as I proceed, two important facts which lead to this conclusion.

For one thing, all the underdeveloped countries are located in the tropical and subtropical zones, while all industrialized countries are to be found in the temperate zones. By its direct and indirect effects on human beings, animals, soil, and various materials, climate is generally a serious impediment to development, although one can read hundreds of books and articles on the development problems of these countries without finding any reference to climate as little as, for instance, to the "soft state" and corruption.

In regard to pollution and to a certain extent depletion

of resources, particularly the agricultural ones, climate implies that these countries stand more defenseless against destructive forces. I would be prepared to exemplify this thesis over broad fields, although in this brief lecture I have to abstain from doing so.

The second important fact, even much more important than climate, is the population explosion.

I find it a serious defect of the so-called Founex Report on Development and Environment, presented as a document to the United Nations Conference on the Human Environment, that these two major problems of climate and population are not given their due importance. It is, unfortunately, not true that the problems of environment in underdeveloped countries are simply related to underdevelopment on the one hand, and development on the other.

VIII

In one sense, the population increase is the key factor in the environmental problem. Natural resources have to be considered in relation to the size of the population which shall be provided for. And pollution is also in many ways a function of the density of population.

The world population may already have reached the 3 billion mark in 1960. It is estimated that it will increase by another billion by 1975 and come to amount to 7 billion at the end of the century.

If, at present time, the underdeveloped countries may be estimated to have two-thirds of the world's population, by the end of the century their population might have increased to four-fifths. At that time the population in the developed countries may have moved toward a stationary condition, while almost certainly the population in under-developed countries will continue to increase further—tending to result in a world population of soon 10 and gradually 15 billion and more. During the seventies, at least, the population increase in the underdeveloped countries is bound to accelerate.

This exercise in extremely loose estimates has only the purpose of conveying a broad perspective of trends in the population field. The driving force behind this development is largely what happens in underdeveloped countries. There the breaking distance is so long, even if birth control should come to be widespread. This is so because of the youthfulness of the population, which is a result of earlier and still persistent high fertility.

And generally the spread of birth control to the masses in these countries is such an exceedingly difficult task. It will not happen spontaneously, as in the developed countries. I cannot enlarge upon this subject in this brief lecture but have to restrict myself to two crucial assertions.

First, that the clamor now so often expressed, not least in the recent debate on the impending environmental crisis, that policy should be very "strict," sometimes explained to imply "compulsory rationing of births" or something in that direction, is totally unrealistic.

What a government, which has decided to do its utmost to spread birth control among the masses, has to attempt is to reach out in the villages and the city slums where people are often illiterate, very poor, underfed and often not in good health, and as a result generally apathetic. There it has to get the individual couples to radically change their most intimate sexual behavior. That is one area where "strict" orders from above do not have much effect and where "compulsion" is simply not workable.

Secondly, there is very little the developed countries can

do to assist governments in the underdeveloped countries in carrying out this policy task, other than giving contraceptives, which they mostly should be able to produce themselves, some equipment and jeeps to be used by the family planning staff, etc. The common clamor that "we" should place a high priority in our aid policy on family planning, demonstrates again lack of realism. There is only one major aid we can offer, *viz.* that our scientists perfect even more the already vastly improved technology of birth control.

IX

That the unprecedentedly rapid population increase in underdeveloped countries, which is going on and will proceed for decades and in any case much more than a generation ahead, places a heavy impediment in the way to development, is clear. Though this is not the place for entering more deeply into that problem I must mention as another serious impediment to development the trend in most underdeveloped countries toward increasing inequalities.

The growth of the population has by itself in many ways the effect of increasing inequalities, particularly in agriculture. But in addition to this, there are gross inequalities that are ordinarily not effectively being counteracted by reforms regarding land ownership and tenancy, the school system as inherited from colonial times, the assessment and collection of taxes, and many other aspects of the institutional structure. In a "soft state" it is also, in the first place, people having economic, social, and political power who can enrich themselves by unlawful means of various types, among them plain corruption that almost everywhere seems to be on the increase.

Egalitarian reforms in these and other respects are not felt to be in the short-term interests of the upper strata which mostly has the power in these countries, whatever their system of government. As in colonial times it is also inevitable that both governments and businesses in developed countries have, through all their relations, come to support these strata, which have not been enthusiastic toward egalitarian reforms but have stood against them or distorted them to favor the not-so-poor.

I have, through my studies, come to the conclusion that the trend to greater inequality must be broken, and that egalitarian reforms in the fields alluded to are not only in the interest of greater justice but a precondition for sustained and rapid growth. Up till now, whatever development most of the underdeveloped countries have had has mostly enriched the tiny upper strata, including the so-called "middle-class" of "educated" and occasionally also the small group of workers in modern industry, but left the masses of people where they were. That the Founex Report is entirely silent on these and related problems is understandable and typical.

This trend to greater inequality in most underdeveloped countries and the certainty in regard to them all of a very rapid increase of the labor force and the entire population render it probable that we will have to foresee an unfortunate development in the seventies, which the United Nations General Assembly has courageously named the Second Development Decade: increasing under-utilization of labor and as a result great misery among the rapidly swelling masses in the rural and urban slums.

In some parts of the underdeveloped world we might then not be far off from the crisis point, when for certain broad strata the Malthusian checks come into active operation. Except after natural catastrophes, such a development will ordinarily stretch over a period of years with minor ups and downs. And it may not for a long time show up in the mortality statistics, which with the great weaknesses of statistics on illnesses is commonly also used as a measurement of morbidity.

But a large and growing part of the poorest strata in a population may be diseased, or at least lacking in normal vigor, and become even more so inflicted, while rates of mortality are still decreasing, due to the cheap and powerful medical tools made available after the war. They might continue to live and breed only to suffer debilitating conditions of ill health to an ever greater extent. But sooner or later the death rate is also going to rise.

X

The environmental problem, particularly in regard to the aspect of depletion of irreplaceable resources, is nowadays discussed as a global problem. But there is a distributional issue involved: who has the power over the resources? The disregard of this issue makes many of the now common brave and broad pronouncements utterly superficial and misleading, indeed meaningless. To give meaning, concreteness, and relevance to our pronouncements on the global problem of resource depletion, we have to lay down as an assumption, needed for drawing our inferences, a definite condition in regard to the distributional issue.

It is customary to say that the 20 or 30 percent of the part of mankind living in the developed countries now for their own use dispose of some 80 percent or more of the world's natural resources. Much of this 80 percent of the world's resources is imported from the underdeveloped countries where now two-thirds and soon a very much greater portion of all people live. I am quoting these loose estimates as they roughly illustrate a fundamental element of inequality in the world today.

International trade in resources and products near the raw material stage has in some respects rather sinister consequences. As my compatriot, Professor Georg Borgström, has done public enlightenment a service by reiterating, underdeveloped countries are continually exporting large quantities of high-quality, protein-rich food products to make overeating possible in the affluent developed countries, and sometimes to provide food for dogs and other domestic animals, or to be used as fertilizers. Thus fish meat, for instance, is imported from African and Latin American areas, more critically short of protein than even South Asia. In the United States and Europe it is then used to feed the broilers and livestock.

The result is that the inhabitants of the rich countries take an altogether disproportionate share of protective food available in the world and use it in a less economical way than would be necessary in the underdeveloped countries, at the same time as they generally use up an equally disproportionate share of grains for feeding purposes.

But even aside from that type of export from underdeveloped countries which deprives them of primary products from land and sea that they would very badly need themselves for feeding their largely undernourished and malnourished population which is now rapidly increasing, the fact must be spelled out that the small minority of people in developed countries appropriate and use for their own production and consumption an entirely disproportionate and steadily increasing part of the world's resources. One broad inference is that any hope that the living levels in underdeveloped countries would ever even approach those in developed countries would presuppose a radical increase in their use of irreplaceable resources.

Particularly with the now-growing awareness of their threatening depletion this would, in turn, necessitate acceptance of a substantial lowering of living levels in developed countries. I see no sign of such a thought, even among the most ardent advocates of the necessity of taking a global view of the use of resources, and certainly not in the announced aid policies of any developed country.

My main point, however, is the purely logical request that any discussion of threatening depletion of resources in global terms, if it shall not remain on a level of general and unclear phrase-mongering, must define a stand on the distributional issue. Is the assumption made that in the interest of great equality in the world there should come to be a more fair distribution of resources between developed and underdeveloped countries in order to make possible a corresponding speeding up of their development? Or is instead the assumption made that the present proportion appropriated by developed countries is going to be upheld and even gradually increased with their rising levels of living, including the large imports of resources from the underdeveloped countries?

The second alternative of status quo on the distributional issue is apparently taken for granted. This should then in all honesty be stated. And the word "global" should be used with more care, spelling out that tacit assumption.

As I have already hinted, there is also a tremendously wide difference in timing, when in particular the race between agricultural resources and population size may result in a catastrophic situation. In some underdeveloped countries that time might now be approaching for large masses of their inhabitants, while developed countries would even be in the position to continue to decrease the areas of cultivated land as well as the labor force employed in cultivating it. A simple counterposing of population and arable land for the world as a whole is a meaningless and misleading exercise.

In regard to most other resources and, in particular, all sorts of minerals, the main efforts to economize their use must take place in the developed countries, where most of them are actually being used up. Any success in doing this will then, for many of these materials which are imported from some underdeveloped countries, imply a decrease of export incomes for these countries: lower exports and, whatever prices are established by price policies in the importing developed countries themselves in order to bring down their use, lower prices in the "world market," that is in this case the export prices for the underdeveloped countries depending on these exports. Again, a "global" analysis in aggregate terms of total world resources and the use made of them is grossly misleading and, in fact, meaningless.

XI

I will now end my lecture by asking: what are the political prospects for governments applying policy measures to improve environment or, to begin with, to prevent its further deterioration?

I am not rising to futuristic heights, least of all trying to press my estimates into quantitative terms, for which there is no basis, and stretch them over decades ahead. I am considering present trends and their possible or probable continuation in the near future, say the next ten years.

Even with the tasks so humbly perceived, I raise no

claim to infallibility. From experience and study of how politics evolves, I have learned that people's actions and reactions, collectively as well as individually, will always be more or less unexpected, and the more so the further ahead we gaze. As I have said: history is not a predetermined destiny.

I will be thinking first of national communities in the developed Western countries, but shall later consider briefly other countries and also international relations.

The awareness of threatening dangers to our environment has in the last decade or so spread to all persons in these countries who are at all alert. Recently this awareness has become intensified, partly under the influence of the preparations for the United Nations Conference on the Human Environment. Under the leadership of its Secretary General, Mr. Maurice F. Strong, this preparation has, in fact, been turned into a worldwide educational movement.

In many of our countries we have seen this reflected as almost a competition among the political parties to urge government action for the protection of environment. We should not hide from ourselves, however, that what I have called the ideology and psychology of unrestrained economic growth has meanwhile retained its hold over people's minds as powerfully as ever. And in national communities with a competitive market economy, every single group is bent upon, and organized for, continual pressure for getting their incomes and their levels of living raised, with little or no visible intention of changing the direction of their consumption demands.

There are, in particular, few if any widespread thoughts on the costs and restraints that are implied in the application of the new policies, and no great preparedness in any group to participate in paying these costs and submitting to these restraints. Such inconsistency in popular conceptions of policy issues is nothing extraordinary but rather a universal pattern of the way politics is conducted in our type of countries.

Meanwhile, however, we already see governments sponsoring studies and setting up agencies for environmental controls. I would foresee that without too much resistance we everywhere in the Western world shall have ever more effective controls over all sorts of novel chemical and biotic concoctions that for some time have increasingly gone into the supplies offered by pharmacies and are being used in food production as cosmetics or to improve taste. Their number is rapidly increasing, and up till now no more than a minor fraction of them are said to have been reliably tested for possible toxicity. The same is true of the increasing variety of washing detergents. The capacity of the national research institutions is increased by borrowing from what is done in other countries. These new discoveries are not kept secret as are often the results of research and development work in industry and, in particular, in the military field.

The restrictive controls over these types of manufactured substances will be the more readily accepted by the public because of the steady stream of new research reports on the damaging effects of one after another of them. I can foresee as soon possible legislation for laying the burden of proving them harmless upon the producer or seller. They form too small a group to be able to offer much resistance as a pressure group. And the restrictive controls do not cause many real sacrifices and costs on the part of the consumers, who might even soon learn that they are better off with a less variegated supply of drugs, fancy foods, and washing detergents.

When, however, we come to restricting the use of various chemicals for raising productivity in agriculture and forestry, and to preventing the dumping of unprocessed waste from industrial factories into lakes and rivers and their poisoning the air by smoke, then we enter a field where substantial costs are implied. The more effectively the controls are applied, the more the costs are increased. If not absorbed by government subsidies and borne by the taxpayers, these costs will ultimately fall on the people as consumers in the form of higher prices and on income earners, mainly the employees, in lower wages than otherwise they could press for.

Only transitionally can these costs be expected to remain losses on already invested capital. But considering the fact that such strong vested interests are involved, it must be astonishing that these controls have recently been increased and magnified so relatively fast in many developed countries. Part of the explanation is undoubtedly the publicity about, and people's actual experiences of, for instance, dead waters where it is not safe to swim and where the fishes are disappearing or have become so contaminated that they cannot be eaten without health risk. In regard to these types of controls we are, however, only in the beginning of what we, with good reason, should continue. And the controls will be expensive.

In advanced welfare states like Sweden, interest will increasingly become directed upon improving the working milieu in the factories in order to protect the health and happiness of the workers. This movement becomes coordinated with the strivings for "industrial democracy" giving the employees a larger say in directing the industrial activity, particularly, but not only, in regard to working conditions. Industry will cooperate, partly under compulsion of government regulations, sometimes sweetened by subsidies in various forms, and partly, undoubtedly, because

92

of social responsibility. But it all carries heavy costs.

In these countries there will also be increasing popular pressure for protection of nature, flora and fauna in order to serve the pleasure and well-being of the people. Nor is this without its costs.

All the policy measures here exemplified are directed against pollution in the widest sense of that term. In a world where raw materials can be imported and are continually cheap, the threatening depletion of irreplaceable resources will hardly become an effective motivation for national policies.

Exceptions to this rule are the land and energy resources, in regard to which restrictive planning is usually also required in order to avoid or decrease pollution. Some saving of resources will also follow the reclamation and recycling that is often implied in preventing wastage from industry or human conglomerations from pestering our environment.

Besides entailing costs, policy measures being taken to preserve and improve our environment will regularly restrain people's freedom to do what they please. This is what I referred to as the administrative problem. In so far as the policy measures are restricted to charges and/or subsidies of a generalized nature, the administrative problem should not be too heavy a burden to shoulder for a government of a developed country, although the sometimes high profits to be gained by circumventing these controls may create temptations similar to those causing tax avoidance and tax evasion. General regulations restraining and redirecting investment and production in industry and commerce and the municipalities's ways of dealing with wastage will also usually not raise insurmountable administrative difficulties.

But when it comes to regulating individual behavior of

ordinary people and, to some extent, also of the really small-scale enterprises, for instance, in what manner they dispose of all sorts of refuse and how they generally behave toward nature, there are limits as to what the government can do both in the costs implied in policing the regulations and in the acceptance by those being controlled. The limits on the practicability of exerting authority to discipline the behavior of masses of people are already now severely hampering policy.

XII

The primary difficulty is, however, the political one: how to get people to permit the government to initiate the policy measures in the first place.

Take a special case: the automobile. Our nations will probably accept ever more effective controls of the production of automobiles, aimed at making them safer and less polluting. The increased costs, ultimately to be borne by the consumers, will in the generally inflationary climate of all developed countries, and with the high priority given to that type of private consumption, be tolerated without too much grudging.

But this is only part of the problem. All big cities are severely overcrowded by automobiles. Not only is the air polluted by their exhausts but the transport situation is in a mess, no government or municipal authority having felt it possible to restrict effectively the use of automobiles in cities. The owners and would-be-owners are everywhere by far the biggest political party.

And in no developed country, as far as I know, has it proved politically possible to get the car owners to pay the full costs—including the heavy investment costs for roads

94

and for the adjustment of cities to the cramming of cars in the streets, the costs impiled in the delays caused to people in the cars and on the streets, the costs for policing the traffic, and the very heavy public and private costs caused by the accidents, not to speak of now paying also for the pollution of the air.

General declarations in favor of "a new style of life," directed upon "the quality of living" while giving up consumption of a lot of basically less necessary commodities, has a general appeal to any public. But the accustomed "style of life" has a great power to survive, particularly in a competitive market economy, where every group is bent upon defending and raising its incomes and levels of living. Almost our entire institutional structure and our attitudes are geared to growth of the old kind. And now comes the additional fact that there are expenses implied in the policy measures needed for defending and improving environment. No group is very willing to pay these expenses.

It is true that in the longer run these costs may be profitable and result in higher productivity, but initially they are heavy. The costs come first, far ahead of the returns. They are in the nature of "investments."

We might at this point note that none of the developed countries has managed to prevent inflation. From one crucial point of view the fundamental cause of inflation is that people are not prepared to make sacrifices in their private consumption large enough to pay for the public expenditures they want to have made.

The priorities differ. For the forced extra saving implied in inflation, the Americans have in exchange a huge military establishment and the deeply disturbing memory of the lost Vietnam war, while we Swedes have, among many other good things, a great number of large and excellently equipped but very expensive hospitals. But fundamentally there is a similarity in the causation of inflation. What the ordinary citizen in our countries wants is higher public expenditures without having to pay for them, that is, he wants to make investments without committing himself to savings.

It is in this situation that we have to take into account that policy measures against pollution imply costs, often of the investment type. Without a fundamental change in people's attitudes, these policy measures will, therefore, add to the forces driving us to inflation. From one angle, important to the economist, this is the effect of people's desire to have a "higher quality of life" but without any infringement on all their other desires. In this development, any single country has to watch what is happening in other countries. If it goes faster than in those countries with which it trades it incurs dangers for its trade and payment balance.

This relation to inflation will undoubtedly put a hamper on the speed with which policy measures against pollution can be decided upon and applied. And if we should make the assumption, at present rather unrealistic, that a country would become prepared to fight inflation effectively and keep an unchanged value of its currency, it would have to go further than otherwise would be necessary in its restriction of private consumption if it wanted to pursue, at the same time, an effective antipollution policy—or otherwise it would have to give up that policy more or less completely.

XIII

Concerning conditions in the centrally planned economies of the developed Communist countries in Europe, I restrict myself to the assertion that the problem is largely a similar one, except that those countries are not under pressure to care so much about their peoples's reactions. This difference is, however, only relative.

As everybody knows, the articulate, educated citizens of underdeveloped countries, who represent their "public opinion," show less excitement about the environmental problems. The oligarchies that are mostly ruling these countries, often not truly representing even those upper strata, are also often much more tightly controlled by the big enterprises, who therefore represent comparatively much stronger pressure groups. Among them the thought is rather natural that their industrialization should not now be hampered by controls that were not applied in the developed countries when they industrialized.

In agriculture, in particular, where there is acutely both depletion and pollution on a large scale, policies to prevent a deterioration of the environment are intimately related to other needs for radical reforms, in regard not only to population control but also to landownership and education, among others. As these reforms are ordinarily not attacked with much vigor, environmental reforms also tend to fall under the table.

It is under the influence of the discussion in developed countries that the problem is now broached even in underdeveloped countries. It is then for natural reasons often directed against certain possibilities that developed countries will inaugurate policies to defend themselves against pollution reaching them through imports. To this I will come back.

XIV

There are intergovernmental problems related to the preservation and improvement of the environment. Let me first consider these problems as they appear within a political bloc of developed countries that, because more like-minded, can be assumed to offer the greatest opportunities for intergovernmental cooperation.

I am here in the position to cite an important and widely noted policy statement by Dr. Sicco Mansholt, President of the EEC Commission.[3]

It is, indeed, radical in its tenets. He argues for "a strictly planned economy" for the Community of West European states: a "European plan," that is supposed to make the planning "highly decentralized" and become "respected when national economic plans are drawn up."

For this planning, the GNP shall be "abandoned" and replaced by a mystical "GNU (utility)." This latter concept is not defined, except by a reference to "Tinbergen's idea of Gross National Happiness," with the added reservation that "it is still not known whether one can quantify this utility." As I have already observed, it is not only a question of quantification: even conceptually this idea is not tenable.

The purpose of this strict planning is to reach "a non-polluting system of production and the creation of a recycled economy." Particularly because conservation of resources is also a goal, Mansholt foresees "considerable reductions in the consumption of material goods per inhabitant, to be compensated for by the extension of less tangible goods (social forethought, intellectual expansion, organization of leisure and recreational activities, etc.)." As there will then be "a sharp reduction in material well-being per inhabitant," Mansholt also finds reasons for egalitarian reforms aimed "to offer equal chances to all."

Mansholt argues "that the Commission could make concrete proposals" for this overall centralized planning

[3] I am quoting Mansholt from the *Manchester Guardian,* April 11, 1972.

and apparently also believes that those will be accepted by the member states.

I am an old planner, with the roots of my thinking firmly in Enlightenment philosophy and, in particular, the thinking of the early socialists in France and England, whom Marx later called "utopian." But experience and study have taught me the very narrow limits for effective planning and plan-implementation in our type of national communities. Even stopping inflation and restoring a stable value to our currencies seems at present an unrealistic goal for planning. And we may note that economists generally are not even pursuing seriously that goal any longer but are satisfied by pointing to the danger of getting out of step with the general trend to inflation in other countries.

That Mansholt is building a castle in the air should be clear from what I pointed out in the last sections. Peoples and governments do not become different because of joining in the community of states, the bureaucracy of which Mansholt is heading. Rather the commitment to seek common solutions of problems will necessitate considerations being given to the government that is slowest to move and thereby often hinders other governments from proceeding faster and further.

This will not prevent the EEC from occasionally being able to provide the matrix for intergovernmental negotiations within their subregion, directed, for example, toward getting the West German government to take action against the poisoning of our common atmosphere by the Ruhr industries.

In regard more particularly to planned policy actions against depletion of resources, which figures so prominently in Mansholt's philosophical tract, I doubt whether any government action is likely, or possible, to be taken in any country as long as there is a world market for these resources—with the reservation made in the next to last section for a few of them where controls are needed already to decrease pollution.

The possibility of pressing a common central planning of pollution controls upon the European Communist countries participating in the Warsaw Pact should be somewhat easier because they are all planned economies, though even in that subregion it will meet resistance, because between underdeveloped countries regional cooperation is generally not highly developed. Moreover, since the pressure for environmental controls is weak within the mall, I cannot foresee good prospects for any such cooperation.

XV

Coming then to the still broader problem of worldwide cooperation to protect and improve our environment, the prospects do not appear bright. We have behind us a number of conspicuous failures to reach intergovernmental agreements in fields where common interests should be very strong and of even more urgent character.

In regard to the negotiations on armament controls, which the United Nations Charter placed high on the agenda for the new world organization, we have so far reached only cosmetic agreements which have not stopped the armament race nor affected the competitive arming of underdeveloped countries, mainly by the two superpowers. The results of the SALT talks, which we are waiting to hear when the text is drafted, can unfortunately not be expected to stop effectively the armament race, but only slightly to redirect it. The two superpowers will have reasons of their own to play up the talks. They have already followed in a strange collusion, for instance, in regard to the partial test ban, which as we know has not hindered increases in underground testing.

Even while, so far, the terror balance between these superpowers has prevented outbreak of open warfare between them and between developed countries generally, military aggression by one of the superpowers—as in Indochina where it, besides other horrors, represents an unprecedented gross destruction of the environment for very poor people—and armed conflicts within or between underdeveloped countries are continually flaring up, and the parties to the conflicts are regularly being aided or even spurred by developed countries, particularly by the superpowers. The machinery for prevention or peaceful settlements of conflicts in the United Nations is increasingly becoming bypassed and rendered obsolete.

Concerning these serious failures of intergovernmental cooperation I can refer to the publications of the Stockholm International Peace Research Institute (SIPRI). Its third comprehensive *Yearbook* will be published and available in a few days.

Globally, the aid to development of underdeveloped countries, which never amounted to much, has in the past decade been decreasing quantitatively; its quality is also deteriorating. This serious development is largely hidden to the general public by a gross falsification of the statistics as published by the governments in most developed countries and by their organization OECD and its Development Assistance Committee (DAC).[4]

Commercial policies in developed countries have in various ways been discriminating against exports from un-

[4] Myrdal, *The Challenge of World Poverty. A World Anti-Poverty Program in Outline,* Pantheon Books, New York 1970, Chapters 10 and 11.

derdeveloped countries and are still doing so on a large scale. An effective way of aiding them would be to end these discriminatory policies and, indeed, generally to make commercial policies work in their favor. It can be proved that such a change of commercial policies in the long run would be in the interest also of the developed countries themselves.[5]

The third meeting of UNCTAD, which has just ended, has not demonstrated much willingness on the part of most developed countries to move in this direction in regard to commercial policies, nor is there any indication of their preparedness to raise the quantity and quality of financial aid.

All of the above, as well as the very precarious situation in the underdeveloped countries which I have alluded to in earlier sections, forms the political climate in the world for the United Nations Conference on the Human Environment, which is now meeting here in Stockholm. I believe it is prudent that we should feel happy, if the Conference can preserve the momentum in the awakening of interest in the environmental problem, set up a permanent agency for continuing the work, build a substructure for carrying it out in the regions and subregions under UN auspices, and, in addition, perhaps outline a few badly needed treaties in regard to intergovernmental cooperation in preventing pollution of air and water.

In the pollution field, however, intergovernmental action will be severely hampered by collision of interests, particularly those between developed and underdeveloped countries. To take one example, if the former countries are stamping out the use of D.D.T. and other similar harmful agents, they might find it natural to set up trade barriers against imports of food and other commodities containing them.

[5] _Ibid.,_ Chapter 9.

As long as science and technology have not produced an effective and cheap substitute for D.D.T. that is decomposed more fully and rapidly, underdeveloped countries may find that they cannot afford to abstain from using it. If that becomes their policy, the developed countries cannot, in fact, protect themselves very effectively since D.D.T. also spreads through air and water. Developed countries, for their part, can, however, feel the need to set up trade barriers for protectionist reasons.

This is meant only to exemplify, by an abstract example, how interests are crossing each other in the pollution field. This does not imply that it should not be possible in many cases to find compromise solutions advantageous to all parties.

But none of the broad pollution problems is simple. For reasons already alluded to, pollution does not stand out as a major problem to those who are politically powerful in most underdeveloped countries. To obtain their cooperation they will have to be offered inducements.

In regard to depletion of irreplaceable resources I must confess that, as long as they are flowing in international trade, I cannot see any possibility of even approaching policy agreements in this field. As I said, in the individual developed countries I cannot foresee any interest in taking action against depletion, even nationally, except in very limited fields, and, of course, action against depletion is even less likely in the underdeveloped countries.

And I see no political mechanism through which action could be taken for preserving resources. We have not a world government, still less a world government with the power to enforce planning on a world scale of the use of resources.

What we have are agreed matrixes for government cooperation. This would do, perhaps, but only if they were used more effectively for intergovernmental agreements on important issues such as those now concerning the United Nations Conference on the Human Environment.

XVI

As I am ending my lecture I realize that what I have said adds up to what is usually characterized as a "pessimistic" view. Personally, I am against both "pessimism" and "optimism," which to me only represent differently directed biases. As a scientist, I want simply to be realistic.

Let me add that when I have had to express rather somber views, I have not fallen into defeatism. I have been taking the shorter view of the immediate future. If we are given time, we might be able to change political attitudes and even political conditions. A realistic analysis should only urge us to strive the harder and to be prepared for pressing radical reforms.

Concerning the environmental problem, it does matter that pending dangers are drawn to everyone's attention. People are generally not entirely cynical, though they are ignorant, shortsighted, and narrow-minded. This implies that to some extent they can be brought to act against what they have become accustomed to feel are their own short-term interests.

Nationally this implies being really prepared to accept a "new style of life." We all feel hopeful about the youth now growing up in the developed countries, that they shall have other ideals than those now steering our course of policies. We also hope that they shall have more compassion for the misery in underdeveloped countries.

Internationally it means that we could become prepared to permit the intergovernmental organizations within the United Nations system to work more effectively as matrices for government cooperation. This also would require the abstention from pressing for short-term and most often narrow and even misconceived interests. It is clear that internationally that would demand greater generosity on the side of people in the developed countries. From that moral issue we cannot escape. Economics itself is a moral science, which, in principle, was recognized as such by our predecessors a hundred and two hundred years ago but is now often forgotten.

Population
Introduction by Ababai Wadia

Uncontrolled population growth is one aspect of the ecological crisis. While developing countries are tackling it as an integral part of their socio-economic advancement the industrialized countries have to focus on it from the standpoint of preserving the environment.

In fact, it has been pointed out that in proportion, population growth can be considered to be even more of a pressure and a liability in affluent countries for, on a per capita basis, their consumption of resources can be as much as twenty-five times that of poor countries, so that even their lower population growth has important adverse effects. The United States, for instance, with 6 percent of the world's population, consumes about 40 percent of the world's output of raw materials, not counting food.

Obviously, what is needed is to evolve a new philosophy of life and to work for moderate levels of comfort spread over the greater part of the world, rather than to have the extremes of rich and poor which become accentuated as times go on. It may well be, therefore, that concern for the environment, if truly pursued, will prove to be a potent means of eliminating global poverty.

Inescapably, therefore, the environment, development, and population must be regarded as three aspects of one whole and must be kept in harmonious balance.

Ultimately, it is the ethical values that will count, to make man's lives truly rich. We are possibly witnessing today the slow emergence of a new and fundamental ethic based on ecologic equilibrium and harmony, which will help mankind to rediscover his place and purpose in life, and renew the fount of inspired living within him. Such a renewal is being longed for by humans in all parts of the globe.

Population

By CARMEN MIRÓ

To participate in the Stockholm Lecture Series sponsored
by the International Institute for Environmental Affairs
is indeed an honor, but more than that, it constitutes a
challenge. Designed by its organizers to cover fundamental
issues which would provide an important added dimension to the United Nations Conference on the Human
Environment, the Lecture Series inevitably had to include population as one of the topics to be examined. But
the passion with which the debate on population has proceeded in recent times, especially when it centers on the
relationship with environment deterioration, adds a new
ingredient to an already difficult discussion laden with
misinformation, not to speak of mistrust. If in any way
my remarks today can contribute to the efforts of eluci-
dating the means and ways through which the popula-

tion factor can be adequately inserted into a global strategy aimed at achieving a dynamic equilibrium between man and the natural milieu, I shall have answered, at least partially, to the challenge posed.

The United Nations Conference on the Human Environment, as the Secretary General has repeatedly pointed out, constitutes a milestone in the development of international cooperation. It is important not only because on an unprecedented scale it has stirred world interest in basic environmental issues, but also because it has helped to accumulate new knowledge which, among other things, will undoubtedly enrich future discussions on how best to tackle the so-called demographic problem.

In presenting the topic of population in this Series, I do not intend to enter the ongoing polemics between environmentalists and demographers. I believe with Duncan [1] that "a concrete human population exists not in limbo but in an environment." As he points out, by mere occupancy of it, "as well as by exploitation of its resources, a human population modifies its environment to a greater or lesser degree, introducing environmental changes additional to those produced by other organisms, geological processes, and the like."

The recognition of this fact and the acceptance of the existence of what might be called a functional interdependence between environment and population, in no way, though, can lead to the conclusion, so often heard today, that population growth is the determining force in the destruction and deterioration of the human environment, which to many is now threatening the very survival of mankind.

[1] Otis Dudley Duncan. "Human Ecology and Population Studies," in Hauser and Duncan, eds., *The Study of Population*. Part IV, 28. Pages 681–682.

Population growth has for a long time been accused of constituting the main obstacle to the development of the less-developed world. More recently it has been burdened with the responsibility for the depredation of the environment. From this latter assertion has stemmed, as a non sequitur, the feeling that the solution to the problem lies in stopping population growth. Curiously enough, the solution is equally advocated for the developing regions where rates of demographic increase have reached very high levels and where the absolute size of the population is already considered too large, as well as for the developed countries with moderate rates of growth and much smaller population size.

This paper tries to uncover the validity of the accusation leveled against population growth as the fundamental source of ecological disruption. Before attempting this, it gives a very condensed description of how population growth is proceeding today and what, in the present view of demographers, awaits us at the end of this century. It also examines rather briefly the short- or medium-term possibilities of attaining a world stationary population. It ends with some personal reflections as to how those interested in the study of population phenomena, especially demographers, can contribute to our understanding of future possibilities and of the consequences of alternative policy decisions.

It should be stressed that the views presented here are strictly personal and do not necessarily represent those of the organization with which I am associated.

I. Population Growth Today and Prospects to the Year 2000

The most recent UN figures [2] place the population of the world for the year 1972 at almost 3,800 million, of which 71 percent (2,670 million) live in the less-developed regions. The same source estimates [3] that during the present quinquennial period (1970–1975) the annual rate of increase in the population of these latter regions (2.5 percent) would be two and one-half times higher than that of the more developed ones (1.0 percent). While the growth rates are expected to start diminishing in both types of regions by the middle of the 1980s, the differential remains exactly the same toward the end of this century, when the world population would have reached, according to the "medium" variant projections, 6,500 million, and the share of the less-developed regions in this total would have risen to 78 percent. Viewed in absolute numbers the world population would increase during the next twenty-eight years by 2,700 million, of which 2,370 million (88 percent) would be added to the less-developed regions.

Another striking feature of present and anticipated medium-term population growth behavior is that related to the increase of urban settlements. Of the world population it is estimated that close to 1,100 million are living

---

[2] United Nations. "Total Population Estimates for the World, Regions and Countries Each Year, 1950–1985." ESA/P/WP.34. 16 October 1970.

[3] United Nations. "World Population Prospects, 1965–2000 as Assessed in 1968." ESA/P/WP.37. 17 December 1970.

today in cities of 20,000 or more inhabitants,[4] with almost equivalent figures for the more-developed and the less-developed regions. If the UN projections obtain, the urban population so defined would be more than twice as large in the year 2000. The most interesting aspect of the development envisaged is that while in the so-called affluent societies the increment would take the urban population to a size one and one-half times larger than the existing one, that of the less-developed regions would almost treble. It should be further stressed that these figures do not portray in its entirety the trend toward urban agglomeration in the latter regions because there would still be an important fraction of the population concentrated in smaller cities.

The estimates just cited are not, of course, intended as an adequate and accurate description of today's demographic situation nor of its expected evolution in the coming decades. They serve only to underscore that the concern with the swelling size of mankind is indeed unavoidable, whatever the ideology, the profession, or the nationality. The magnitude of the task that lies ahead will require that men show not only ingenuity but also courage and generosity. Ingenuity to devise adequate solutions, courage to apply them, and generosity to do so even if it implies forsaking some earlier anticipated advantages.

II. The Prospects of Attaining a Stationary World Population

Perhaps overwhelmed by the factual evidence advanced by demographers regarding the expected medium-term

[4] United Nations. "Growth of the World's Urban and Rural Population, 1920–2000," in *Population Studies*, N⁰ 44, New York, 1970.

evolution of the world population, even the most ardent proponents of population-growth curtailment do not advocate the attainment of population stabilization, or more properly, of a stationary state before the end of this century.

As pointed out by the authors of *The Limits to Growth*,[5] history has witnessed recurrent proposals for "some sort of nongrowing state for human society" since Plato through Malthus to Boulding.

Furthermore, analytical exercises on the demographic conditions necessary to move in the direction of a stationary population and the demographic consequences accruing therein have long been a favorite subject with formal demographers.[6] But public debate on what is now called "zero population growth" is very recent. As has been the case with several other lively discussions on population matters, it started in the United States. It was prompted by the publication in early 1970 of an exhortation by the United States representative in the UN Population Commission [7] to the effect that his country, and with it the rest of the world, should strive for a zero rate of population growth by the year 2000, at the latest. This

---

[5] Donella H. Meadows *et al. The Limits to Growth.* Universe Books, New York, 1972.

[6] See, for example, P. Vincent. "Potential d'accroissement d'un population," in *Journal de la Societé de Statistique de Paris,* January–February 1945; Tomas Frejka. "Reflections on the Demographic Conditions Needed to Establish a U.S. Stationary Population Growth," in *Population Studies,* Volume XXII № 3, November 1968, pages 379–387; and United Nations. "The Concept of a Stable Population," in *Population Studies* № 39, New York, 1968.

[7] William H. Draper Jr. "Is Zero Population Growth the Answer?" An address at a testimonial dinner in Washington, D.C., 2 December 1969. Published by the *Population Crisis Committee.*

was followed by discussion of the topic in the 1970 meeting of the Population Association of America,[8] when Notestein branded the proposal a "platitude" though recognizing its value as a possible "organizing focus for research and educational efforts concerning the importance of a world-wide trend to stationary population and the means by which it is ultimately to be achieved."

This is precisely what it has turned out to be. Bourgeois-Pichat,[9] using the population of Mexico as the basis for his analyses, tried to uncover what would be the true meaning, in demographic terms, of attaining a stationary state by the year 2000 and retaining it afterward. The U.S. Commission on Population Growth and the American Future dealt with the subject as "a principal issue in its deliberations." [10] As part of the research undertaken for the Commission, Coale [11] prepared a paper on "Alternative Paths to a Stationary Population" in which he examined the feasibility of attaining and maintaining a stationary population in the United States. The Club of Rome's Project on the Predicament of Mankind [12] is also an exercise on how to attain a "state of global equilibrium," including in this, of course, the stabilization of population growth. The Latin American Demographic Center (CELADE) is presently exploring the resulting demographic

[8] Population Index. Vol. 36, N° 4. October–December 1970, pages 444–465.

[9] Jean Bourgeois-Pichat. "Un taux d'accroissement nul pour les pays en voie de développement en l'an 2000 rêve ou réalité?" *Population,* 25 année, N° 5, 1970, pages 957–974.

[10] "Population Stabilization." Chapter 10 of the Report of the Commission.

[11] Ansley Coale. "Alternative Paths to a Stationary Population." Paper submitted to the 1971 Meeting of the Population Association of America.

[12] See footnote 5.

consequences for some Latin American countries of pursuing the goal of a stationary population. No doubt many other demographic research institutions are studying the topic.

Up to now, all the research recently undertaken in this area leads to the conclusion—indeed not very comforting to the advocates of zero population growth—that given the present demographic conditions, the goal is not reasonably attainable within the next three decades, much less, of course, immediately.

Even assuming that political, ethical, and ideological considerations would not constitute formidable obstacles to the adoption of a world population stabilization policy,[13] the built-in potential for growth present in most of the world population for reasons of its age structure makes it impossible to attain stabilization without profound demographic disturbances which no population would stand, not even in the face of coercion. Just for illustrative purposes Bourgeois-Pichat found that to attain stabilization by the year 2000 and retain it, fertility in Mexico would have to fluctuate from 0.6 birth per woman in 1995–2000 to 3.8 in 2040–2045, when measures would have to be taken to reduce fertility again. In the same manner the age structure would show tremendous oscillations. These can be appreciated more clearly in certain functional age groups. The school age population, for example, would grow from 4.7 million in 1960 to 7.2 in 1975 to decrease to 2.0 million in 2010 when it would start growing again until it reached 6.2 million in 2055. Similar fluctuations would occur in the economically active ages and in the population over age sixty-five.

[13] For a rather poignant examination of these aspects, see the comments made by Paul Demeny as discussant of Notestein's paper on ZPG, published in Population Index (see footnote 8).

The U.S. Commission on Population Growth and the American Future arrived at the same conclusion when searching for "criteria for paths to stabilization." In this regard it stated that

> While there are a variety of paths to ultimate stabilization, none of the feasible paths would reach it immediately. Our past rapid growth has given us so many young couples that, even if they merely replaced themselves, the number of births would still rise for several years before leveling off. To produce the births consistent with immediate zero growth, they would have to limit their childbearing to an average of only about one child. In a few years, there would be only half as many children as there are now. This would have disruptive effects on the school system and subsequently on the number of persons entering the labor forces. Thereafter, a constant total population could be maintained only if this small generation in turn had two children and their grandchildren had nearly three children on the average. And then the process would again have to reverse, so that the overall effect for many years would be that of an accordion-like continuous expansion and contraction.[14]

After these considerations, while the Commission chose not to recommend any specific path, it has shown in the Report that an "optimal path" would be that leading to stabilization fifty years hence, that is to say, in 2020, when the U.S. population would have reached 330 million.

With a younger age structure and a higher level of fertility, both producing a larger growth potential, it is easy to conclude that the ZPG goal would entail much more difficulties in the less-developed areas.

[14] "Population Stabilization." Chapter 10 of the Report of the Commission.

Now, if the prospects under present demographic conditions are for continued growth and the possibilities of attaining its stabilization seem somewhat removed in time, we have to reach the inevitable conclusion that the world will have to accommodate many millions more. It is precisely at this point that environmentalists claim that mankind is faced with its gravest predicament because population growth accelerates the destruction of the human environment, the depletion of which, they say, has already reached levels endangering the continued survival of man on this planet. With no intention of inviting complacency in dealing with future population growth, I postulate that it still remains to be shown how much and in what ways the depredation to which the human environment has been and continues to be subjected can be attributed in the less-developed areas mainly to the sheer increase in the numbers of its occupants. Clarifying this would indeed contribute to charting the way to a sounder policy of environment conservation.

III. Man as Predator of His Environment

Whether one chooses to define human environment in the rather restricted sense of traditional ecology or with the more ample meaning now being attached to it, one has to accept that as far as history can record, man has plundered the earth in every conceivable way and that it is not at all easy to find a direct relationship between the increase in size of a population and the pace at which destruction and depletion of the environment has proceeded. In fact, history tells of utter mismanagement of the natural milieu when population size, density, and rate of growth were considerably lower as compared with present-day levels. The classical example frequently cited is that of Mesopotamia, where an ancient civilization flourished having decayed later, mainly because of unforeseen consequences of the agricultural exploitation: irrigated lands had to be laid to waste due to their salinization and irrigation canals had to be abandoned due to silt accumulation.[15] The inability of the State to cope with the demands placed upon it for the conservation of the land, while at the same time protecting the nation from invaders and internal conflicts, is the reason most frequently cited for the degradation of the agricultural lands. Population size and density seem to have played only a minor role.

But we do not have to resort to history in search of evidence that other factors are indeed much more important than population size and growth in precipitating negative impacts on the environment. In the less-developed areas the destruction, depletion, and deterioration of the human environment have been underway through many diverse forms and long before modern pollutants would have created the alarm on which the present crusade "to save the environment" is based.

The following remarks, while primarily based on Latin America, are in more or less degree generally applicable to other areas.

Outside of the cities, the indigenous population, frequently sparsely settled in the territory, is an active agent of environment destruction. The continuous clearing of lands for subsistence shifting agriculture, with its indiscriminate destruction of forests, has long been an important source of ecological imbalances. These practices,

---

[15] J. H. Stallings. _Soil Use and Improvement._ Prentice–Hall, Inc., Englewood Cliffs, N.J., 1957.

though, have little to do with the size and the rate of population growth. They are more directly related to the social organization, especially land tenure systems and levels of education of the agricultural producers.

There is no denying that an increased number of subsistence farmers might accelerate the destruction of woods and forests, but, even at the risk of sounding platitudinous, one has to say that reducing their fertility would accomplish very little in the direction of protecting the environment. It might very well be that an increasing density of this type of farmer would contribute to reducing aggression against the ecosystems. Ester Boserup [16] postulates that "sustained demographic growth among primitive people does not always result in deterioration of the environment, because the possibility exists that the population, when it outgrows the carrying capacity of the land with the existing subsistence technology, changes over to another subsistence system with a higher carrying capacity." In other words, in this situation, which has actually existed in backward areas of Latin America, Africa and Asia, demographic trends have acted as an adapting factor, forcing developments which otherwise might have not occurred.

A measure which would indeed contribute to bettering the lives of the subsistence farmers, even under present social organization, while at the same time protecting the forests and with them all the other resources (climate, soils, water deposits, etc.) so dependent on them, would be a drastic attack on the massive felling of trees for commercial purposes, which much more rapidly and much more completely has already irreversibly destroyed innumerable portions of the territory of the less-developed

[16] Ester Boserup. "Environment, Population and Technology in Primitive Societies." Unedited article.

areas. This spoilation started long before the so-called population explosion and threatens to continue unless strong coercive measures are applied.

Overgrazing is also blamed for soil deterioration, with its concomitant influences upon environment, but very seldom are the rapidly increasing number of peasants the owners of the herds causing the destruction. It is a well-known fact that the cattle raisers, usually belonging to the most favored strata of society and because of that contributing in lesser degree to the high level of fertility, occupy the lowlands, drawing the subsistence farming to the highlands with the consequences described earlier. Paradoxically, these lowlands have been more rapidly destroyed in the face of the decreasing human population density being replaced by a growing cattle density.

Some of the evils just mentioned refer to what might be called traditional agriculture. The introduction of modern agricultural technology, highly dependent on chemicals, is a relatively recent addition to the threats to environment in the less-developed areas. Not unrelated to the need of feeding a growing world population, it is only indirectly connected with the increasing size of the native populations which continue to be underfed, while certain developed nations pay their farmers for maintaining idle vast extensions of agricultural land.

Similar remarks as those just made would apply *mutatis mutandi* to the exploitation of aquatic resources. Aside from the ill effects caused in rivers and seas by soil degradation and forest destruction there are the added consequences of man's fishing activities. Here again some catching methods, such as the use of explosives, are extremely detrimental to the conservation of certain species. But much more harmful is the over-fishing practiced with large fleets and "modern techniques." It could be argued that these activities have been carried out in such manner to respond to the increasing demands of a growing population. It could probably be easily demonstrated that the reasons for the use of such environment destructive practices are more related to increased financial benefits than to population growth *per se*.

If we would turn to the exploitation of the other natural resources, especially minerals, usually undertaken in the less-developed areas by foreign enterprises, we would find that also here the human environment has been the subject of massive and, in many cases, irreparable damage. The extraction of minerals which, like bauxite, require the prior removal of the top soil, and the selective extraction of high-grade ores, are only two examples of land degredation. These activities are also an important source of water pollution, with all its attendant consequences. An important portion of the industrial production based on these minerals is being diverted to increase the per-capita consumption levels in the highly affluent societies and not necessarily used to respond to the demands of the growing number of inhabitants of the less-developed regions.

The prime mover for the concern about environment has been the rapidly increasing tempo of urbanization, both in the more-developed as well as in the less-developed areas. In the former regions what appears to be at stake is the "quality of life"; in the latter, the quest is just for mere "survival," for the larger fraction of the urban population. For the rest, the fight is against the perils of industrialization. In other words, the urban settlements of the developing regions suffer the evils of development without receiving an equivalent fraction of its benefits. Herein lies the greatest difficulty of turning conservation of the environment into an appealing goal to the majority of the urban masses. When they have to struggle for adequate

122

water supply; when their health is under continuous threat due to the absence of the most elementary sewage and waste disposal systems, for lack of safe methods of food manipulation or because of the invasion of rats and other equally pernicious animals; when they suffer from unemployment and are sick, undernourished, ill-fed and ill-dressed, it is indeed almost impossible for them to comprehend why resources which could be used to better their lot, should now be channeled to clean the air and purify the water.

There is no doubt, though, that urbanization—more a problem of population distribution than of population growth—is creating very critical problems all over the world. Agglomeration in the cities, with its sequelae of marginal unhealthy settlements at their peripheries, is causing, among other things, the rapid destruction of agricultural lands which were once the source of agricultural produce consumed by the city dwellers. The high agricultural productivity levels prevailing in the more-developed countries, coupled with the fact that they are importers of food items, reduces the negative impact of this process. The voracious spreading of city boundaries in these countries, gives ground to the concern, much more than anything else, for the disappearance of leisure and recreation areas. In the less-developed countries the continuous physical expansion of cities, besides adding to the problems of providing badly needed infrastructure, imposes further limitations to the possibilities of a balanced urban-rural development. In summary, any campaign addressed at promoting the conservation of the environment in urban areas of the less-deveolped countries would have to be based on the search for solutions which would maximize the enjoyment of the benefits of industrialization, while minimizing its perils.

The preceding discussion was not intended as an exhaustive examination of all possible sources of environment destruction which are only indirectly related to population growth. No mention has been made, for example, of the lethal effects on people and environment as a result of massive bombing and the use of chemicals in warfare which has now reached unprecedented levels. The cases described here are drawn mostly as illustrations of the risks involved in taking the simplistic position of attributing to the sheer increase in numbers the main responsibility for the recently discovered environmental crisis. The predatory nature of the "homo œconomicus" more than his reproductive urge seems to account for his suicidal efforts in subduing the environment beyond reason. Forms of social organization, land tenure systems, levels of literacy, national and international industrial and trading practices, to mention only few, are important intervening factors which merit extensive consideration. Appealing to the need of population control as a means of environment conservation without accompanying it with an equally strong plea for drastic measures to change the social and economic conditions which have made possible its massive destruction, depletion, and deterioration could evoke suspicions that the less fortunate inhabitants of this planet are being confronted with a new Malthusian argument.

IV. The Potential Contribution of Demographers to a Viable Future

Approaching the subject of environment conservation paying due attention to its many ramifications, in no way should be construed as implying that nothing should be done regarding population growth and distribution.

Neither demographic increase nor concentration in cities can go on forever. "Spontaneous" reversal of these trends is not to be expected. Therefore policies are being applied in an attempt to modify them and new more effective measures would have to be devised. It is here that demographic surveillance and research could play a very valuable role. The early detection of new population trends; the guidance in the collection of needed data; the interpretation of the demographic impact of alternative policies; the projection of expected population evolution; the feasibility of attaining specific demographic goals and the measuring of the attendant consequences; the elucidation of the interrelations of demographic variables with other social and economic factors, and the explanation of past demographic behavior as a means of understanding that prevailing today and of chartering future potential evolution; these, and many others, are tasks being developed today by demographers the world over. These activities will necessarily have to be incremented in the coming years if demography, as a science, is to contribute to the harnessing of the forces impeding the access to well-being and happiness to the majority of mankind. The forthcoming UN World Population Conference should serve to focus population in its right perspective vis-à-vis economic development and environment conservation.

Science, Technology, and Environmental Management
Introduction by Chief Adebo

Some of the forecasts of environmental danger have been so scarifying that a number of people have simply refused to believe them, but there is no doubt of the good they have done. We have been shocked into recognizing that there *is* a danger, that the danger is very real, and that remedial action is called for.

As a national of a developing country, I am glad that it is now generally recognized on our side that pollution of the environment is not simply a problem of the affluent, and that it is generally recognized on the part of developed countries that, while the protection of the environment is a common interest of all, the measures to be taken in that regard, and in particular the apportionment of the burden flowing therefrom, must take careful account of the development needs of developing countries.

Introduction by George Baranescu

Between environment and technology there exists an obvious contradiction, because technology is the force that compels the environment toward ends which would not be achieved in the natural course of events. Some of these technological transformations lead to the intended objective, which is to serve people. But technological processes can also lead to other kinds of transformations and that is the field where contradiction between man and technology can arise.

Unfortunately we do not know the limits of man's adaptability to diverse situations because we cannot yet describe the evolution of life in relationship to the environment. In this regard one cannot rely too heavily on past experience because conditions are not the same.

However skillful may be our technologies, however innovative may be the responses of science, and however disposed governments may be to apply them, we shall obtain complete and decisive success over the long run only if everyone on earth is profoundly convinced that to work for the preservation of his own environment is also to accomplish a noble duty in behalf of generations to come.

Science, Technology, and Environmental Management

By LORD ZUCKERMAN

The Stockholm Conference is taking place in an atmosphere of confused concern about the possible—or impossible—state of man's future environment. The extent to which agriculture and industry have polluted our physical environment continues to cause misgiving. We worry about the rapid rate of increase in world population. And we are fearful lest we may be exhausting the world's nonrenewable resources and, because of modern methods of production, endangering the future of some of those which are renewable.

These are real fears and real problems. But at the outset of my address, and like others who have already spoken in the context of this Conference, I want to refer to another, and in my view far more urgent concern—that our worries about the physical environment of tomorrow may be helping to distract attention from some of the more immediate problems of mankind—problems such as the poverty, illiteracy and ill-health which still afflict the greater part of our world, and those conflicts of interest between nations which keep erupting into war. The recent agreements between the Federal Republic of Germany and the USSR, and between the United States and the USSR, have helped to reduce some of the tensions by which the world is afflicted. But they are only a beginning. Much still needs to be done before we can all sleep safely in our beds in the knowledge that, whatever else, nations will have the time and resources with which to improve the human environment of tomorrow.

I want to make another general observation in introduction. Because of modern means of communication, every part of the world is conscious of what is happening elsewhere. The world of today is much more literate—in quality, if not in quantity—than it was little more than a quarter of a century ago, at the end of the Second World War. In spite of the poverty which still exists, standards of living have in general improved since then. But the consciousness of the disparity in living standards between and within countries has also made us impatient to better our lots, and impatience is leading to a worldwide sense of frustration. One of its manifestations is the increasing physical and intellectual violence to which we are all subjected. Another is a sense of hopelessness that we shall never be able to solve our social, political, and environmental problems. Hence, in my view, the growing volume

130

of protest that in our greed we have allowed science and technology to run away with us, and that our material civilization is leading us inevitably and rapidly to our doom.

There would, of course, be no science if there were not at the same time a sense of hope that answers can always be sought, and usually found, to properly formulated scientific questions. Equally, with the world full of problems, there would in my view be far less reason for hope than there may be if there were not the promise of more science and technology. But it is not just because I am a scientist that I find it difficult to believe that we no longer control science and technology, and that we are controlled by their merciless force. It is because current discussions about the impact of science and technology on the environment seem so often to denigrate the good they have done. So although the topic to which I have been asked to address myself in this series of lectures is "Science, Technology and Environmental Management," let me begin by referring to some of the benefits the world owes over the past twenty-five years to the impetus of science. The years have not been entirely black.

First, I would remind you of the spur which science gave throughout the world to the postwar expansion of education. The state of education still leaves much to be desired, but it would certainly be worse than it is if a major shortage of scientifically and technically trained manpower had not been revealed when it came to dealing with the problems of postwar reconstruction, and with the needs of the emergent lesser-developed countries. In the United Kingdom and other Western European countries —and I believe in the United States as well—this shortage was the major stimulus which resulted in the postwar growth of universities and schools. The uninterrupted emphasis on scientific education in the USSR could perhaps be regarded as an extension of a prewar recognition that knowledge of the sciences is a necessary condition of economic growth and social advancement.

That the growth of education in universities and schools has not been confined to the sciences does not affect my proposition. The point I am making is that the great postwar revolution in education was "triggered," not by the cry that there was a critical shortage of historians or philosophers, but by the message that countries were in desperate need of more scientists and engineers. A further point to remember is that the machinery for mass education over the past twenty-five years is the product of applied science and technology. Without radio and TV, the recognition of man's present predicament would certainly not be as widespread as it is now, nor indeed would the enlightenment which leads to understanding—and often to protest.

After education, I would put the enormous benefits which men, women, and children owe to the work of medical scientists. I am not going to worry you with statistics. The fact is that the major killing and disabling diseases which used to plague the world—malaria, tuberculosis, poliomyelitis, typhoid, cholera, and the simple fevers of childhood like measles and diphtheria—have been eliminated over the past two to three decades in many countries. Given the means, they could be eliminated in others. Antibiotics and analgesics and a host of other drugs have also enormously lightened the burden of illness to the individual. In those sectors of disease where such developments have had their greatest impact, they have also considerably reduced the direct and indirect cost of ill-health to the state in those countries which enjoy the benefits of effective medical services. I said I would not quote any statistics. But I must refer to the fantastic changes which the past two decades have witnessed in the average expectation of life—and in every country which has been touched by public health measures and by the advances of modern medicine. In countries like the United States, Sweden, and the United Kingdom, the increase in the average expectation of life has affected every age group, and most people in these countries already expect to live longer than the biblical limit of three score years and ten. In the poorer countries, the impact of the increased expectation of life has been most marked in the younger age groups, and most significantly in the numbers of people who reach reproductive age. The influence the sharp fall in death rates, unaccompanied by any significant decline in fertility rates, has had on the size of population in underdeveloped countries is in all our minds.

If the chances of leading a healthy life have improved because of the advances of medical science, it is equally true that had it not been for science, living conditions in the world as a whole would, in general, have been considerably worse today than they are. It is only too easy to deplore the poverty, hunger, and misery which still afflict hundreds of millions of the world's population. But these things would have been immeasurably worse if scientific knowledge had not been put to work in order to alleviate them over the postwar years.

There *has* been social progress in our time. For all the daunting social problems with which the world is wrestling, can one contend that the conditions of life for the majority are not better today than they were, say, fifty years ago? Hunger and poverty persist, but surely not on the scale which the world has known in the past. For all their continuing defects, living conditions are surely better now than they were in the years before the war. The developed countries may still be short of houses, but they have more houses, and more houses with baths, running water and indoor sanitation. And more and more people now enjoy the new dimension of personal liberty which the motor-car confers.

In what I say here, I am not thinking only of the developed countries. In some lesser-developed countries there has also been a striking improvement, especially over the past ten years. Rome was not built in a day; and ten, twenty, even fifty years, are a mere flash in the life of nations. The improvements that can already be observed in some lesser-developed countries are surely a foretaste of more.

In short, it would be idle to pretend that, in helping to transform the world in the postwar period, science and technology have not brought major and wanted benefits to vast numbers, and potentially to all. But the human appetite and human expectations inevitably grow as fast as they are satisfied, and now they are growing faster than the possibilities of satisfying them. The inevitable has happened—reaction in some quarters is going all the way to the denial of the value of industrial growth, or even of the merits of the Green Revolution.

Our newspapers, urged on by a plethora of pseudoscientific books, articles, and speeches, are filled with items which warn us that irreversible damage is being done to our physical environment, that the day is not far distant when it will be impossible to satisfy our energy demands, leave alone our need for raw materials like metals. We are also warned that nothing is likely to happen which will stop human population growing until the world ends with "standing room only"—to cite the title of a book which was published in the early thirties! All these fears have now been concentrated in a small book called *The*

Limits to Growth which was edited by a young computer specialist at MIT and published under the auspices of the Club of Rome. The message of this book is that unless we mend our ways; unless we bring about an abrupt change in our industrial processes and in our patterns of consumption; and unless we stop breeding at more than replacement rate, our society will come to a harsh end—and within a period which would only be a minute fraction of the time it has taken it to evolve. This is what is now envisaged as the "Predicament of Man."

Before I move on to the topic of how science and technology can be used in the management of our physical environment, let me cite three main reasons why we pursue science. The first is that our intellectual curiosity drives us to discover more about the world around us and about ourselves—in modern jargon, about "what makes us tick." The second is that we want to improve the way we manage our social systems. The third is that we need scientific and technological knowledge to help us control and manipulate our environment.

It is strange that, at a time when no more than a fraction of a percent of mankind can be said to be living in conditions which denote that they have some mastery over their environment, some people should be contending that science and technology are moving too fast. These writers seem to have grave misgivings about man's increasing ability to provide the material things he wants with less expenditure of labor of a disagreeable and life-shortening kind.

I am ready to believe that some science has been a menace. In medieval Europe, a man who fell ill was probably at least as likely to survive if he did not find himself in the hands of those who then made up the medical profession—the physicians, the apothecaries, and the barbers.

The same is probably true of some lesser-developed countries today. But primitive medicine was never in any real sense a science; it represented and represents the application of scientific ignorance, not knowledge.

Only time will tell how—and not whether—the resources problem solves itself in relation to the growing demand of the world's swarming, multiplying, and deprived millions. But the environmental issue is surely something we can and should get into perspective lest future generations condemn us for the actions which could derive from a heartless extremism.

We all know that in the past mining and factories and sometimes even agriculture played havoc with the natural countryside—that is to say, played havoc in our eyes, not presumably those of our grandparents who allowed it to happen—and that as industrial cities invaded agricultural land, the working-class houses which were built, and which at first were a vast improvement on the hovels they replaced, in due course became the slums of today. We all know that our rivers became polluted and that some even became open sewers. We know that our air became fouled with dust and SO_2 and even HCL.

And above all, we know that the position would have become intolerable if measures against the undesired effects of industrialization had not been taken, and could still become intolerable if they are not pressed forward in future.

But who are the scientists, who are the technologists, who are the rational people who would have the process of environmental spoilation continue? Surely we have begun to learn our lesson? If we had not, would the UN have called the Conference under whose umbrella we are meeting? And surely we have learned that the social costs of industrial and agricultural progress are not charges on the future which can be permitted to accumulate endlessly. We recognize that account must be taken of the true costs of the production whose fruits we, as consumers, wish to enjoy now. In the end, and wherever it occurs, man will have to pay for the environmental damage he has caused.

There are some extremists—professional scientists not among them, but men who comment from the sidelines, and not on the basis of practical experience—who see pollution as a menace which must inevitably grow. I, however, know of no scientific evidence for this view, whereas I know from my own experience that devastation of the landscape can be corrected, that rivers can be cleaned, and that the skies can be cleared.

Where today are the once-notorious pea-soup fogs of London—the way visitors enquire about them now one would think they had been one of England's tourist attractions—and how is it that the Thames, once all but devoid of fish all the way from its estuary to well past London, now boasts more than fifty species of fish? The Thames is only one of the rivers of our small, densely populated and industrialized country which is in process of being cleaned.

The worst enemies of the environment, as well as of mankind, are poverty and ignorance, and the big lesson we have learned is that all environmental improvement depends upon scientific and technical progress. These create both the knowledge of how to avoid pollution and the wealth needed to avoid it. People sometimes speak as if the British Clean Air Act of 1956 was the result of a sudden realization that smog can be lethal to people with bronchial illnesses. The deaths in the London smog of 1952 were the immediate stimulus for this Clean Air Act, but the basic causes were that we had learned how to have clean air and that the time had come when we could afford what it cost. In the twenties and the thirties, prohibiting the domestic coal fire, which is what has produced the improvement we see today in England, would have been economically impossible. Even the working man in regular employment could not have afforded to use any fuel but coal, still less could the widows, the pensioners, and the unemployed. The change has come from improvements in fuel technology, from a fall in the cost of oil, and above all, from the growth of wealth which has come from the advance in technology generally.

Pollution is essentially a social and political problem. Its scientific aspects are relatively simple to deal with. The real problem is how much should we pay to reduce pollution at the cost of not doing other things which are also socially and politically essential? The concept of "zero-tolerance" for pollutants is meaningless socially and scientifically, and ruinous economically—as Dr. Stokinger, the Chairman of the U.S. National Research Council's Committee on Toxicology has put it in a recent article.

I have referred to a book *The Limits to Growth,* which has been hailed—mainly by the scientifically uninitiated —as a scientific statement about man's environmental problems. Its authors led themselves through the circuits of a computer to the conclusion that the only way out for mankind is to slow down economic growth abruptly and to change human nature drastically. We have to alter our social and political institutions so that we behave more sparingly than we do with our raw materials and also so that we divide our industrial product more equitably than we do today. If we do not do these things, we shall be digging our own graves.

This conclusion derives from a theoretical analysis of the likely effects of an exponential growth of human populations, of an assumed exponential use of resources, and of an assumed but inevitable increase in environmental

138

pollution. For my part, I have no hesitation in saying that I am among those professional students of our environmental problems who dismiss the book as unscientific nonsense. None of the basic assumptions of the book bears much relation to the empirical evidence we have about any of these matters. But—and here is my main criticism, for I shall not add to the host of others that have been leveled by professional writers—the only kind of exponential growth with which the book does not deal, and which I for one believe to be a fact, is that of the growth of human knowledge, and of the increase in the kind of understanding with which we can imbue our efforts as we see to it that our increasing numbers do not become incompatible with a better life, and with a physical environment which is not just a reflection of what the doom-watchers deplore in what they see around them now. For example, if it were even partially agreed scientifically that the use of supersonic civil transport could wreck the ozone layer which overlies our atmosphere, can we seriously imagine that we would not find ways of inhibiting the use of such an aircraft as our knowledge of their secondary effects—if any—became more apparent? What are we— ants, lemmings, or rational human beings?

Obviously man's power to do harm in this age of nuclear weapons has reached planetary proportions. But the alarm which we now experience in fact comes largely from our increased knowledge of the risks we have always been running and of those we are still running. The danger of being poisoned today by lead is probably as small as at any time since lead started to be mined. But we have now developed the ability to detect the presence of this element, and of others like mercury, in very small concentrations. This enables us to take countermeasures where they are called for, and—what is not altogether unimportant—to abstain from taking blunderbuss precautions through not knowing enough to be selective. Scientists have, it seems to me, not only a duty to warn governments and their fellow citizens of damage they see occurring or can predict; they also have a duty to correct what they believe to be misconceptions, even though these may be propagated by people whose ideals they may share.

Having said all this, how am I to interpret my theme of the relation of science and technology to environmental management? Let me begin by explaining briefly what I personally understand by science and by technology.

To me, science is a body of knowledge to which more is always being added, at the same time as what is already accepted is in constant danger of being eclipsed by new observation or by the results of new experiment. I believe, too, that most practicing scientists see science as consisting of those disciplines which depend on controlled and verifiable observations and experiment, which are subject to constant check, and the results of which may permit inferences that can lead to general laws. Whatever else, arbitrary speculations about the environment which are based on misstatements of publicly ascertainable fact, cannot be dignified as scientific knowledge.

Science, moreover, is not just a search for new knowledge. It also comprehends those activities which harness or apply basic scientific knowledge, the engineering and the many technologies which create wealth through industry and agriculture. To the working scientist, to the man whose career is in one or other branch of the biological or physical sciences, the attempt to separate basic or applied, as though the various aspects of science evolved in some special order, can only lead to confusion when one tries to understand the troubled phase through which the world is now passing.

For science is above all one of the most powerful of the factors—perhaps today the most decisive—which determine the structure of our society, and which act on the body of our beliefs, on our understanding, and on our aspirations. In this sense, science is no more the servant of the democracies than of the authoritarian regimes poised against them, and of the philosophical and political concepts on which they, in turn, are based. In both kinds of regime, we find nations which are rich, and nations which are poor; nations which already enjoy the fruits of highly developed industrial economies, and nations whose economies are underdeveloped; nations which are favored by sophisticated educational systems, and nations which are in large part illiterate. Everywhere—both in our own democratic and in the authoritarian worlds—we find the same urge for greater personal fulfillment, for higher standards of living, for more speedy economic advancement, coupled only too often with an overriding frustration, which at the national level translates itself into a sense of insecurity deriving from the clash of interests in the international arena. All sides in the present world struggle will use science and technology where and how they can in order to survive and in order to achieve their respective national aims.

We all live in villages, towns, or cities. These—to use a much abused word—are the kernel of man's ecosystem. We should be in major trouble if science and technology did not provide us with assured and safe supplies of water, with adequate sewage systems, and, above all, with the food that is conveyed to us in manageable and preservable form from the country. We need power to provide us with heat and light, as well as to drive the machines of industry. We need timber and metals and synthetic substances like plastics with which to build the apparatus of our social existence. We started on this course of social evolution back in the Neolithic period, some 10,000 years ago, and we cannot turn back now. But we are beginning to be conscious of what is entailed in the maintenance of a human environment in which populations are increasingly urbanized, in which people live in tall buildings serviced by electric power, and in which we depend on trains, planes, and motorcars. One might say that science and technology have, in that apt idiom of speech, "painted us into a corner." The United States, with 6 percent of the world's population, now consumes some 35 percent of the energy output of the world and some 40 percent of its processed physical resources. It has willy-nilly established a target for human aspirations, and aspirations which we know are unlikely to be satisfied with the world divided as it is, and with nations stratified in varying layers of material privilege. Industrialization has established itself as the one sure cure for poverty in a world the bulk of whose population still lives by subsistence farming; and history as yet gives no examples of any but small communities which have voluntarily turned their backs on higher material standards of living. Instead, a uniformity of desire and demand is generated for the so-called good things of life as the one world discovers how the other lives, and what it lacks itself. Obviously, we cannot say that the economic history of the West will be recapitulated as industrialization spreads, and as the chains of the past are broken in distant parts of the world. But we can be all but certain that neither the needy nor the rich will allow the process of applying the fruits of scientific knowledge to stop, either in the national or in the international frame. In this process, means become ends, because as new ways of doing things are discovered, they transform the

142

things being done, and so their purpose. That is why I say science and technology, by determining the pattern of human society and of our material evolution, have "painted us into a corner." Now that we are aware of this fact, if we wish to prevent the corner becoming smaller, we have to use more science and technology. We have to use more science and technology to achieve further industrial and agricultural growth in order to deal with the social and environmental problems with which we are faced.

The UN Conference on the Human Environment has been convened to deal with some of these problems. It has met at a moment when all responsible people recognize, however vaguely, that there is an environmental problem, or several problems. What industrialist today does not believe that it is the duty of government to give orders about pollution and the duty of the citizen to comply? The attitude of nations has changed similarly. I shall give one example. Last June, representatives of countries bordering the North Sea met for an informal discussion about the dumping of wastes at sea. By the following March, a convention about dumping had been signed, and governments pledged themselves to pass legislation to effect it as a matter of urgency. But this was not a case of reacting to damage already caused. Catches of fish in the waters in question have never been higher. A danger was foreseen, and governments cooperated promptly to prevent it. And this is the essence of the use of science and technology in the management of our physical environment.

But here I feel compelled to repeat something that I have already said. I do not believe that catastrophic pollution of the planet is among the worst risks that mankind now faces. In Great Britain, pollution is not increasing. In spite of growth in the population and the continuing

143

growth of our economy, our air is becoming purer, our land is becoming more fertile, and our rivers are running cleaner. We are far from being alone in this experience.

There has been foresight. To the best of my knowledge, it is only rarely that some disaster affecting human health has occurred before it was recognized that pollution was occurring. Obviously disasters have occurred; for example, typhoid and cholera have in the past been spread either through polluted water supplies or carriers, and on rare occasions, food supplies have been affected by, for example, botulinus toxin. But in all of these cases what was lacking was effective scientific knowledge—not foresight.

British environmental laws have in general been based upon the preservation of amenity, on laws which forbid the preventable pollution of the atmosphere or the devastation of a landscape. In the course of time we have learned that some kinds of pollution are measurable and at the same time reversible. Air and water pollution fall into this category. It is because we now have the resources—as I have said—that the air of the United Kingdom is pure and its rivers in process of regeneration. Many years before Rachel Carson's book *The Silent Spring*, we also imposed strict regulations on the use of certain types of agro-chemical. I remember this only too well, because I was chairman of the committee which the government set up to deal with the question.

There are, however, some other forms of pollution which can be measured, but which are in practice irreversible, and which for that reason have to be prevented. Here I would cite as illustration radioactive waste. There is yet another kind of pollution which is not measurable in objective terms, and which might or might not be reversible—namely, the changes which inevitably occur in our physical environment as human populations grow

144

and, with them, our towns, industries, and roads. From the very day that organized human village settlement occurred some 10,000 years or so ago, part of what was then natural landscape became transformed. Today, there are very few acres in the United Kingdom which represent the natural landscape as it was before man started to spread, and particularly into those parts of the globe which are most densely populated.

I have heard it said that if this process goes on, man will bring about the destruction of those animal species which still remain, in the same way as he eliminated the dinosaur. But man never destroyed the dinosaur. Long before human beings roamed the earth, vast numbers of animal and plant species were eliminated in the process of natural selection.

We obviously cannot restore the landscape of the Triassic or of the early Pleistocene. But for the first time we know what it is we are doing as our influence spreads over the globe, and so long as our species survives, we can consciously take the necessary steps to preserve to the best extent possible the fauna and flora which we are determined to have preserved.

In doing this, we must, however, remember that our own values about what constitutes a desirable landscape will not necessarily be those of the generations following us, any more than we accept the values of those by whom we were preceded. What we have to decide is the kind of environmental amenity which we can afford to safeguard in the face of all the other demands made on our resources, particularly those which will be demanded as we try to alleviate the poverty, the sickness and the illiteracy by which we are still surrounded.

The idea that a stationary state of human economy would have to follow a period of economic growth because of a scarcity of resources, population pressure, and falling profits, is as old as the Industrial Revolution itself, and its formulation certainly required no modern computers. At whatever moment one might choose in the two hundred and fifty years since economic growth through industry began, one would have had to predict impending world disaster if the then existing rate of growth were plotted exponentially against known supplies of current materials, and of possible substitutes within the framework of existing technology. But world disaster has not occurred—because new science and technology are not as limited as some assume them to be. Whatever computers may say about the future, there is nothing in the past which gives any credence whatever to the view that human ingenuity cannot in time circumvent material human difficulties. The parameters which determine a continually evolving human society are not the rigid ones of an engineer's machine shop. We ran out of wood to make the charcoal necessary for the production of iron, supplies of which were for a time threatened. We then discovered how to use coke for smelting, and I do not suppose it will end there.

Serious questions are, of course, now being asked in various countries about their future supplies of energy. Ever since the emergence of nuclear power, hopes have been expressed that here we have a limitless source of energy with which to satisfy man's needs as our fossil fuels dwindle. But fears have also been expressed that we shall not be able to deal with the dangerous radioactive waste products of this source of power. Let me say straightaway that I do not accept this view. There are at least as many competent scientists who deny as affirm this proposition. As always in social and environmental matters, we are dealing here with a balance of risk. Of course there are dangers from nuclear radiation; who would deny them? There are also environmental dangers from the burning of coal and from the building of dams for pumped water storage. But the risks from radiation are understood far better, as has recently been pointed out, than the majority of hazards which mankind has to face. Let me quote from a statement recently made by two professors of physics in the University of Birmingham:

> Even if we assume that there is no threshold below which there is no danger, the deaths which could at most be caused by the effluent of the present British nuclear power programme would eventually build up to about two people per century. A complete replacement by nuclear power of the 70 million tons of coal which Britain uses each year to produce electricity would increase this number to perhaps one every seven years.
>
> In comparison, pollution of the air by coal and oil burning now is responsible for the death from bronchitis of about one person every hour (one a day due to power stations alone). As for genetic effects, we have no accurate knowledge of the number of mutations produced by sulphur dioxide, vehicle exhausts and so on, because little research has been done on this, but it certainly must be far greater than the one per year which would be expected to result from an electricity system based entirely on nuclear power.

I am no nuclear physicist, and I am sure that some scientists in relevant fields might qualify what I have just quoted. But I am enough of a scientist to be guided by men who know rather than by pseudoscientists who weave their fantasies out of the fears expressed by a minority.

I do not want to be misunderstood. There is undoubtedly a qualitative difference between mankind's power of doing damage to his habitat a couple of generations ago, and the power which he possesses today, and still more the power he will have in future. Until recently, the worst we could do was gradually to turn a few hundred square miles of fertile land into a desert. It is only recently that we have acquired the power to do damage on a global scale. And there is certainly no guarantee that our knowledge of the consequences of our actions, and still less our political will to exercise restraint, will always grow more vigorously than will our power to do damage. All I am saying is that it is unscientific to speak as if technological advance is working inexorably toward making a worse world and, in particular, that we are going to be unable to deal with our waste products.

There is one urgent category of scientific problem which needs to be solved or agreed as we face some of the environmental risks which the knowledge we now have allows us to define. That is the problem of deciding wise standards of pollution and environmental damage. Leaving aside the question of amenities, in which I include noise as well as visual amenity, we have, I believe, to set environmental standards in relation to the realities of health, as opposed to the increasing sensitivity of the instruments which measure chemical pollutants. Science will never be the means whereby we get people to agree about visual changes in the landscape. But it can certainly be the means through which we all agree about what chemicals are dangerous to health.

Unfortunately, the enthusiasm which is displayed by some antipolluters seems to be in inverse proportion to their knowledge of the scientific facts. We should not allow ourselves to be driven by extravagant and frequently fallacious statements into getting our priorities, which are still far from being perfect, into even worse shape. Various estimates for different countries imply that protecting the environment from further pollution, and repairing the devastation of the past, would cost somewhere between 1 and 2 percent of the gross national product. And it is said that since GNP is growing in most of the countries concerned at more than 1 or 2 percent, the cost could be readily afforded. Various environmental measures that have been introduced over the past few years in the United States or are now under active consideration, add up on average to an annual charge of something like 20 billion dollars.

But however necessary or desirable the measures concerned may be, is this the right order of priority? Twenty billion dollars is about five times the total of all official grant aid from the richer to those poorer countries of the world who are now struggling to improve their lot. It is more than the total of all grant aid, loans, credits, and investment capital flowing to the lesser-developed countries. It is not for me to answer the question I have put. I merely ask whether a climate of opinion highly charged with environmental emotion is necessarily conducive to the wise assessment of national or international priorities about the deployment of resources.

Nothing I have said should leave the impression that I believe that the course on which we are set will infallibly lead to a better world. But I want to make it quite plain that I regard the environmental dangers which we face as far more manageable than I do the social and political problems which exist in a world in which we have so far not discovered how to eliminate war, in a world in which nuclear weapons could finally devastate not only ourselves but our physical environment as well, and in a world in which the disparity between rich and poor is an endless source of tension. As I have said, the physical and measurable problems of the environment are matters with which science and technology are competent to deal. But they cannot deal with the value systems which determine which amenities should be preserved—these belong to the political domain where all citizens have a voice. Nor can science resolve the dilemma of banishing poverty through economic growth on the one hand, and the achievement of agreed environmental goals on the other. That again is a problem for society at large.

But if science and technology are to serve as they should in protecting our physical environment, it is essential that scientists and technologists should be allowed to pursue such of their work as affects environmental problems—in which I include the preservation of fauna and flora as well as the determination of permissible levels of pollution; the scientific study of reproductive physiology as well as that of the factors which affect patterns of fertility; and the enormous range of technological and economic problems involved in what is now called resource management and substitution—in a hopeful and scientific spirit and not in one of hysterical computerized gloom. The tree of knowledge will go on growing endlessly, and, given time, so will the compassion necessary to use its fruit for the benefit of all mankind. New knowledge always has to subject itself to the most rigid discipline of scientific method. But a free world will never tolerate its application in a dirigiste way, in order to achieve so-called environmental purposes. That is something which professional scientists and technologists who are concerned with the problem of the human environment already know, but which some of those who preach on their behalf still have to learn.

Human Settlements
Introduction by Ro Chung-Hyun

Allow me to state my opinions concerning human
settlement and environmental problems in terms of
micro and macro perspective.

- From a micro view we simply endeavor to alleviate the immediate problems of the human
  settlement area, ignoring the validity and in-
  trinsic worth of its social fabric.
- A macro approach, however, endeavors to discover the original cause of environmental problems fundamentally stemming from human settlement, in addition to uncovering solutions to
  such problems in a comprehensive and organic
  manner.

I understand that the Western concept views the
environmental problem, especially that of the low-
income community, negatively, seeing it as a run-
down and hopeless community. In the Asian view,
however, could we not reconceptualize the Asian
slum community as a superficial squatter- and low-
income community which, in its deep substructure,
possesses positively the creative and developmental
potentiality? Such a potential does exist in the slum

community if the dwellers are guided by an intelligent and well-planned policy.

You and I are all too painfully aware that our urban facilities, without exception, are excessively overloaded and cannot anywhere afford an increase in inward migration.

If the policymakers in the developing countries earnestly try to deconcentrate the population from primate cities, then they need to set up balanced regional and national, comprehensive developmental plans and avoid remodeling after the established Western industrial and urban policy. Only then can the developing countries safeguard and protect the environmental quality while they promote their rapid economic development.

Aristotle wrote: "Men come together in cities in order to live. They remain together in order to live the good life."

Human Settlements

By AURELIO PECCEI

Many years back I read a book of fiction in which our
planet was invaded by a species of voracious and, what we
would now call, polluting creatures. Men united in fighting them. These creatures were also cunning and adaptive,
and they imitated human beings in everything. Since they
were also serviceable, in some lands men stopped fighting
them and found it useful to ask them to conduct a human
life, and do human work. But these invaders were also
prolific, and they were assigned one for each man and
woman and child of our own species, so that every person
in the world had his own replica living side by side with
him. Soon the original world inhabitants were in trouble
because of this locust-like swarm of old and new population, and desperate conditions were brought about in
many countries, giving rise to all the scourges of the Apoc-

alypse on a massive scale—famine, pestilence, war and death. And all this for the simple reason—let me put it this way—that in a very short time the number of people in the world had practically doubled.

I have not been fortunate enough to listen to the previous lectures of this series, but certainly many of you have. Such as have can compare notes and wonder whether the nightmarish fiction I have just quoted does not bear a striking resemblance to the situation building up in the real world, and which we will be facing in the next three or four decades, when the earth population will actually double that of today. This may explain why, as the last speaker—last in many ways—I have been asked to talk about *human settlements*. Since the moment I rather foolishly accepted to close this set of lectures—which has seen some of the best thinkers of our time on the rostrum before me—I have been haunted by the vision of 6 or 7 billion people crowding our globe by the year 2000 or thereabouts, and the very human problem of settling them and providing for their needs decently in this terrestrial abode, which is ours as it will be theirs on equal rights.

If we take the long view—as one always should in considering the problems of our collective future—before coming to our theme some general considerations on our species are in order.

Man is a queer animal—an arrogant, difficult, and aggressive one. He stands in a category apart. In strictly evolutionary terms, all living creatures have, as their supreme urge, that of continuing their own species and making it fitter. One can argue that all forms of life that populate the earth now have been able to adapt to all changes that have so far occurred through the centuries and millennia in their environments—including those caused by man's

154

action. They have learned the *ecological rule of survival* that wants the good of the individual to be subordinated to the good of the species. In this sense a tiger truly knows how to be a tiger; a swallow, a swallow; a spider, a spider.

The human species is peculiar in that it complements its slow biological adaptability with a social organization and cultural devices capable of interpreting and modifying its environments. Hence, its ecological fitness and survival depend essentially on continual adaptation of this extraordinary sociocultural endowment to changing conditions. But, since humans possess another peculiarity—that of being themselves the principal agent of change on earth—and since man-provoked change is nowadays revolutionary, their life and future hinge on rapid and radical, continual cultural updating—as is necessary to maintain the human system in dynamic equilibrium and harmony both within and with respect to its environments. Whenever this complex of internal and external balances has been disrupted in man's history, man had to suffer the consequences. Now these disruptions—as everything else in our time—have reached unprecedented dimensions, and occur for the first time at global scale. *This is the predicament of mankind.*

There is little wisdom in denying that the human system is in the grip of a very serious crisis, and that our growth syndrome, if not cured, is going to make this crisis even worse. However, there are still people who do not see, or pretend not to see, the mismatch between human growth geared on exponential curves, as it is now, and the finite nature of our planet. Not to see the changed reality, these die-hard apologists of material growth—generally economists—bury their unidimensional heads and reasoning under any kinds of arguments, even GNP projections; or fight last-ditch battles maintaining, for instance, that, when *"external diseconomies"* are *"internalized"* in the cost structure, the cost-price mechanisms of the marketplace will take care of practically everything that has gone wrong in society and the environment. More modern economic schools recognize that up to about the middle of this century the human race was sufficiently small in relation to the size of the world for it to be reasonable to treat the ecological system—with its bounty of clean air and water, raw materials, and so on—as virtually unlimited; while now the balance between supply and demand is shifting rapidly. Their rationale is that, if no price is paid for things the supply of which is not unlimited, the consequence in time will be a serious misallocation of resources—which is what they maintain has got to be put right.

However, the crisis of our society is bigger and deeper than what this reductionist approach induces us to believe, and certainly requires a much more drastic cure than a reallocation of resources. Our very cultural basis has to be questioned, reassessed, and probably radically modified—for it is *here* that the roots of our crisis lie. *Anthropocentric,* our culture gives man the illusion that he is the pivot of the universe, the salt of the earth, whereas he is but one of the beings therein, sustained and conditioned—as all the others—by the biomass as a whole. *Egocentric,* it makes man, easy to be convinced as he is, believe that in his species individual good makes for that of the collectivity, and that his community is worth at least as much as mankind. *Utilitarian,* when not downright hedonist, it looks to material satisfactions, to immediate dividends, neglecting the permanent philosophical, ethical, and social values. *Analytical,* it pursues a thousand branches and loses sight of the tree, ignoring the forest. *Philatelic,* it collects notions, data, information, without knowing how to recognize their interconnections. *Technological,* it is so carried away by its bravura that it makes man advance without a thought as to his steps, which are by now giant strides toward the unknown. This culture has transformed man's very condition on the planet, giving him a capacity and power that he knows not how to wield. Man, thus becoming master of his destiny, has lost the sense of it. He does not know where he is, whither he wants to go, how to be truly a man.

After this quick overview of modern man's predicament, we begin to perceive how difficult, even desperate, is the problem of his future settlements. This problem may be approached from many viewpoints. One—the most common—is to focus our attention essentially on the combined effects of the twin phenomena of population growth and urbanization, which are expected to remain rampant in the years and decades to come; and deduce that in practically every country we will witness the formation of unprecedented, fantastic, immense conurbations. From this we would infer that from a world in which the population is mainly settled in villages, cities, and a few metropolises, mankind has quickly to move into a world of very large cities, metropolises, and megalopolises. In commenting on this thesis, I will give you only a few figures, because those I have found are generally and widely discrepant. Furthermore, though accepting as an inevitable fact that a very large and increasing proportion of our contemporaries and descendants will opt, or be constrained, to become city-dwellers—mainly swelling ever more the present urban settlements—I am convinced that this forecast of the inevitable formation of colossal megalopolises needs much probing.

The most imaginative, and in my view most optimistic, painter of future scenarios, Herman Kahn, some years ago

157

wrote with Anthony Wiener a few pages about this supposed trend and the human ant-hills in which people are
expected to live, love, and die when half of the world
population is urbanized. He talks about the three Gargan-
tuan megalopolises which will surge up in the United
States along the shores of the East Coast, the West Coast
and the Great Lakes, and which he calls only by acronyms.
One, Boswash, the megalopolis that will extend between
the ancient, namely the present, cities of Boston and
Washington, may well contain some 80 million people, a
quarter of the estimated population of the United States
of that time. The other two are tentatively called Sansan,
spreading between San Francisco and Santa Barbara, and
Chipitts, which is expected to go from Chicago to Pitts-
burgh and to englobe also the Toronto region of Canada.
Along these lines, one is brought to think that similar
monster cities will explode around the present urban areas
of Tokyo, Calcutta, Paris, London, and so on. Some demographers say that Mexico City is expected to grow to
more than 30 million people, and that probably the same
fate awaits São Paulo in Brazil.

The prospect of these colossal human agglomerations
reflects the trend of thought that "bigger is better" and
that whenever something turns wrong science and technology are there, ready to come to our rescue and solve
our ills and problems. In my view this is a totally misleading vision of the shape and quality of our future. But
it is also, I hope, a sobering and highly useful one, for it
will prompt us to do everything in our power to prevent
it from materializing. This horrid rationalization of the
human settlements to come purports that the compaction
it will require of human beings in ever smaller quarters,
but on a mass scale, is acceptable to the majority of them
as a way of life, while I believe that—however weak and

passive—they will eventually insurge and reject it. And this thesis presupposes also that the organization of city life to satisfy the citizens' needs at such high levels of concentration is feasible, while I maintain instead that it is impossible. For, not only would the psychological and behavioral problem of "live and let live" in high-density areas at such a macroscale become insoluble, but material problems would have no solution either. These megalopolises dreamed of for our future would be a hundredfold more complex and impervious to management than our present-day cities, which already baffle us so greatly with their problems of individual and public, urban and interurban, transport and parking, of airways approach, of subsoil utilization for utility networks and facilities, of education and leisure, of crime and drug prevention, of pollution control, of incoming supplies and waste disposal.

Technologists—other unrepentant addicts to growth—can and hopefully will go on suggesting new fixes, devices, equipment, even solutions, but they can address themselves to a minute part only of all these problems, which are problems of a predominant human and social character. Nor can the "industrialization" of city functioning help much, because the scale factor would be too adverse. In factories themselves life is undergoing profound changes. Experience shows that labor-intensive, partially automated, industries, to which cities may be likened, tend to set a limit to their dimensions and to improve the quality of life within their perimeter—otherwise they become unmanageable and ineffective, and the factory community disintegrates.

Although I have no alternative plans to submit at this stage, this is not the crux of the matter. Not only does this simplistic and mechanistic approach to human settlements in the future have to be discarded—for the reasons already mentioned—but we must also realize, as a question of principle, that no fundamental issue of the years and decades to come can be properly understood and dealt with in this piecemeal way—even if each peace is supposed to contain a large chunk of humanity. Only by a *holistic and global* approach can we hope to establish the cognitive and philosophical context within which the issue we want to attack can be meaningfully examined. This change of optics would permit us better to see the nature of our problem, which in reality is that of *settling and servicing 6 or 7 and then even more billion humans* on our small planet. And this optics would also permit us to see that only by acquiring a greater insight into today's human condition and world situation, and the interactions and interdependencies of every problem, event, and action with everything else in the Man-Society-Nature-Technology integrated systems can we hope not to become lost in the modern world complexities.

It is true that proceeding in this way we will not find easy, immediate answers to the questions we now believe are relevant. But we will learn what is more important: that other, deeper, primary questions must first be asked, and that the life-and-death challenges to our future, as well as our greatest opportunities, are those linked with the overall situations and trends in the planet, not those emerging from more limited environments, or sectoral interests.

Let me therefore depart once more from the theme and make some considerations along these lines.

Our planet has been compared to a spaceship—Spaceship Earth. In the mind of many, this is likened to an immense oceanliner, majestically sailing in the cosmos with a large complement of passengers—first-, second-, and other-class passengers, from those extremely well cared for to those whose lot it is to live in the ship's dark hold. Being in Stockholm, we can say that its citizens are certainly first-class passengers, as are also all the participants at this multiple Conference, whether they are from developed or underdeveloped countries, and be they officials or pro-testers, and whatever their genuine or pretexted interest in mankind's welfare and environment. All of us here are among the privileged, who consume most of the provisions available. Some hundreds of millions of people on board our common ship also fare well, some more, some less well than us. But we know that, on the contrary, the vast ma-jority of the passengers are just "marginal men" living in deprivation; and that their basic aspiration and hope is to raise appreciably their standards of life, education, and opportunity. We know also that these are impossible ex-pectations according to the ship's present organization and rules, and begin to realize that competition and strife among classes and groups seriously impair the condition of Spaceship Earth.

Thus, a distinct, although yet unclear, perception is creeping about on Spaceship Earth, that something impor-tant has gone wrong, that there exists a politically, let alone morally, intolerable *state of injustice* and confusion aboard, that the ship's environment is deteriorating at an uncomfortably rapid pace. Affluent passengers who have heretofore been mainly bent on consolidating their privi-leges, now, I guess, realize that these privileges have be-come strident, even valueless in the face of threats looming up in the future. Second-class passengers—accustomed to receive from the upper class benefits, security, and a body of values they were unable to produce themselves—see that they can no longer keep the lower classes at arm's length, now that these latter are resolved to recoup, improving their standards at all costs, even at the cost of permanently damaging Spaceship Earth.

Symptoms and omens are with us that one of the revolutionary reshufflings which now and then, here and there, have shaken in the past the vertical structure of society is going to happen again—this time occurring, however, simultaneously and ubiquitously within and among all human groups. People keep an eye on their weapons and provisions onboard. The mood and mentality on Spaceship Earth may one day change from that of a liner to that of a *lifeboat*.

The city has to remain the highest expression of human association and polity—home, workplace, forum, academy, stadium, and restplace at the same time. To keep it so, the real issue is how to preserve, or better, to restore, wholesome and enriching qualities to its life, and make it the living place of progressive people, independent-minded, freedom-loving citizens, and if you want, of selective electors and responsible taxpayers, but also of poets and painters, of cathedral builders and birdlovers—*of real men*—and not of computerized, regimented, disaffected, alienated, hopeless, or rebellion-prone multitudes. Only if the city remains *"à la dimension de l'homme"* will it be possible to lay the conditions for these qualities to flourish—and have cities at all.

Under these conditions, we must grasp better how quantitative changes at today's high levels may produce qualitative consequences. Physically, the compounded growth of human population and activities will catapult mankind into new orders of magnitude and complexities already within two or three decades. There are even many sectors in which the next doubling in the production-consumption cycle is expected within ten years, and the next doubling in another ten. Tensions beyond any of our past or present experience will develop in the world.

Most of you have heard of the research study made by MIT for the Club of Rome. It is a study of the modern world's dynamics made with the aid of a computerized simulation model analyzing the trends and cross-impacts of five critical phenomena possessing exponential growth tendencies and which were selected as characterizing the situation of our time. They are the growth *of population, of both industrial* and *agricultural production,* and *of pollution,* and *the depletion of nonrenewable natural resources.* The preliminary conclusions reached during this study are condensed in a general report published in a book under the title of *The Limits to Growth.* In a nutshell it warns that, if mankind keeps growing according to the present tendencies and the related forecasts, it will soon saturate the earth, overshoot the physical limits of its supporting capacity, and finally collapse.

As the whole exercise was not intended to be predictive —it is certainly not a piece of futurology—the time and mode of collapse are indicated only in exemplificatory ways. But the invariable answer to all assumptions made was that, if there is not fairly soon a basic change in the main human trends, collapse will indeed occur—a growth-and-collapse behavior being inherent in the dynamics of exponentially growing phenomena, such as those described, when they occur in a finite environment, such as our world.

The main conclusion to be drawn from this study is that *equilibrium* within the human system and between it and its environments will anyhow be re-established. Clearly, it is in our collective interest rationally to plan for it, even at the cost of heretofore unimaginable sacrifices, and not wait for forces beyond our control to settle it—which will probably occur at the cost of tremendous human suffering. On the other hand, *society in equilibrium* does not mean stagnation. Non-material-consuming and non-environment-degrading activities may be pursued indefinitely —such as education, art, music, religion, scientific research, sport, social interactions, and most service activities.

What we have to consider here is the perspective of collapse, because it will render at least dubious the possibility of further human settlement.

Collapse may happen because humans will spread their presence and activity so greatly that forces in the planetary ecosystem will expel or destroy man as a noxious intruder who upsets the everchanging but always harmonious webs of life. Should this happen, it will mean that *homo sapiens,* the last of the big animal species to appear on our globe, would prove to be a mistake in the biosphere's evolutionary process, a mistake that the delicate but stubborn mechanisms of life would move to correct even if this meant them going back a few million years—just a moment in the aeons of time: suffice it to think that the dinosaurs reigned for 200 million years. Life will then start again in another direction, free from the danger of producing another freak of the human type.

Collapse may also occur because man dissipates the non-renewable resources of the planet, which are necessary for the technical society he has engineered. These resources have been accumulated in successive geological ages, during billions of years before the age of man. But man, trapped by his cultural fallacy that he is king on earth— not just a tiny part and parcel of nature—wants all these resources for himself, calling them "the common heritage of mankind." He is, however, inconsistent even with his philosophy, because he squanders and uses up many of these resources within the short span of a few generations, converting them into waste and pollution, thus depriving the next generations of the riches to which they are supposed to have equal title, and leaving them instead the legacy of an unspeakable mess to clean up.

Collapse may also be caused by war and civil strife—if, for instance, the second wave of human population which will invade the planet in the next three or four decades does not find a place to settle or the means to satisfy its needs. This grim alternative has to be considered with all due attention within the scope of our theme.

I have tried to show, examining it in many ways, that the most tormenting and really frightening problem confronting mankind is *the mass of babies we all go on procreating,* and *who will grow and themselves breed* and *pathetically pursue happiness*—and not only demand the means to keep body and soul together which in itself will already be a difficult business.

Even if we focus only on the subject of today's discussion, we must realize how staggering is the macroproblem of *just providing the physical infrastructure* to receive this supplementary wave of human population in our finite, limited, vulnerable, probably already ill and overcrowded planet. The job of building the houses, schools, hospitals, churches, and churchyards, and the roads, ports, bridges, and transport systems, and of erecting the factories and reclaiming the lands required by this population doubling— *a job to be finished in some thirty years*—equals the construction work accomplished by mankind over the last two billenia, even if we are incapable of duplicating Venice and Bruges, the Taj Mahal and the Kremlin.

I challenge the optimists at all costs to consider this *hardware problem*—we all know that the problems of then operating the hardware, and the software problems generally, are much more intricate—and to imagine how it can be met in all its aspects, from the conception of an overall plan to the investments and financial question, from the standardizations and rationalizations required to the organization and management of this unparalleled enterprise, and last but not least the environmental safeguards which are imperative.

The Club of Rome would like to give a contribution to the study of the *clusters of problems* related to the *settlement of* and *catering for* the multibillion people society looming up for the end of this century—when most of you will still be living. For, we—mankind—are faced by a harsh alternative. Either we accept to grow to 6 or 7 billion during the next thirty years, in which case we must prepare right now in order that all of these 6 or 7 billion might actually live on our common earth; or we do not consider this an acceptable outcome of our present doings, and then we must right now change them, and make the sacrifices and take all the enlightened measures for this outcome not to happen.

Considering the next—and hopefully last—world population doubling as inevitable, the Club of Rome has asked Professor Jan Tinbergen, the Nobel laureate, to inspire and lead a research project whose terms of reference, in a capsule, are to explore whether, or better, under what conditions, this multibillion-people society can occupy the earth with due respect to human ecology over time and with more equitable distribution of the total product among all its members. In other words, it will explore *whether and how* economic and ecological systems can be coordinated in a planetary context at these high levels of population, *whether and how* global development is achievable in manners compatible with earth-keeping and social justice. If the answer to these questions is negative, now or in the future, I am afraid that humankind will face the danger of a Darwinian "Battle for the Earth," a scramble among people and nations to secure space, resources, life chances.

If you consider my presentation not totally disheartening, well, it is because you believe, as I do, that we still have some time, though not much time, to change course. Man can be reformed, if he can see why and how, and he usually finds his finest hour in adversity. He must feel the challenge, however, now and understand the total nature of this challenge. He must also know that the cost of his answer will be very dear and that, short of a profound ethical renewal and a new humanism, his future will remain bleak, whatever his power and capacity. And above all he must realize that he needs to be a much better man if he is going to live during the next century.

Index

172

DATE DUE

DEMCO 38-297